METHODS IN FOREST CANOPY RESEARCH

METHODS IN FOREST CANOPY RESEARCH

Margaret (Meg) D. Lowman
Timothy D. Schowalter
Jerry F. Franklin

UNIVERSITY OF CALIFORNIA PRESS

Berkeley   Los Angeles   London

University of California Press, one of the most distinguished university presses in the United States, enriches lives around the world by advancing scholarship in the humanities, social sciences, and natural sciences. Its activities are supported by the UC Press Foundation and by philanthropic contributions from individuals and institutions. For more information, visit www.ucpress.edu.

University of California Press
Berkeley and Los Angeles, California

University of California Press, Ltd.
London, England

Library of Congress Cataloging-in-Publication Data
    Lowman, Margaret.
Methods in forest canopy research / Margaret D. Lowman, Timothy D. Schowalter, Jerry F. Franklin.
        p.   cm.
    Includes bibliographical references and index.
    ISBN 978-0-520-27371-9 (cloth : alk. paper)
    1. Forest canopies—Research—Methodology. I. Schowalter, Timothy Duane, 1952– II. Franklin, Jerry F. III. Title.
SD387.F59L69 2012
577.34072—dc23   2012006827

Manufactured in China
19 18 17 16 15 14 13 12
10 9 8 7 6 5 4 3 2 1

The paper used in this publication meets the minimum requirements of ANSI/NISO Z39.48–1992 (R 2002) (Permanence of Paper).∞

CONTENTS

CONTRIBUTORS TO TEXT BOXES

YVES BASSET
Smithsonian Tropical Research
Institute, Panama
bassety@si.edu

STEPHANIE A. BOHLMAN
University of Florida
sbohlman@ufl.edu

CATHERINE (CAT) CARDELÚS
Colgate University
ccardelus@colgate.edu

SAMANTHA K. CHAPMAN
Villanova University
samantha.chapman@villanova.edu

BRUNO CORBARA
Global Canopy Programme
b.corbara@globalcanopy.org

SOUBADRA DEVY
Ashoka Trust for Research on Ecology
and Environment (ATREE)
soubadra@atree.org

CHRISTOPHER J. FROST
Warnell School of Forest Resources and
Conservation, University of Georgia
jasmonate@gmail.com (or cfrost@uga.edu)

T. GANESH
Ashoka Trust for Research on Ecology and
Environment (ATREE)
tganesh@atree.org

LEON KAGANOVSKIY
Touro College
leonkag@gmail.com

BEVERLY E. LAW
Department of Forest Ecosystems and
Society, Oregon State University
bev.law@oregonstate.edu

MAURICE LEPONCE
Royal Belgian Institute of Natural Sciences
maurice.leponce@naturalsciences.be

SCOTT E. MILLER
National Museum of Natural History,
Smithsonian Institution
millers@si.edu

WILLIAM R. MILLER
Baker University
william.miller@bakeru.edu

ANDREW MITCHELL
Global Canopy Programme
a.mitchell@globalcanopy.org

VOJTECH NOVOTNY
University of South Bohemia
novotny@entu.cas.cz

GEORGE D. WEIBLEN
University of Minnesota
gweiblen@umn.edu

PHILIP WITTMAN
Tree Foundation
drphil@canopyquest.com

PROLOGUE

The call of "timber" rang out in the forest as the feller completed his back cut and the immense old-growth Douglas-fir tree began its fall to the ground. Branches, tops, and all matter of organic material and dust filled the air as the tree fell and then burst into hundreds of pieces as it crashed onto the ground. After material had stopped falling, Dr. William Denison, an observer and professor in the Botany and Plant Pathology Department at Oregon State University, began working his way through the debris, shaking his head and despairing of any possibility of reconstructing either the structure or the biological content of the tree's canopy.

Denison was a key member of the infant research team founded in 1968 to analyze an old-growth Douglas-fir ecosystem at the H. J. Andrews Experimental Forest, deep in the Cascade Range of Oregon. Funding had been provided by the National Science Foundation as a part of the US International Biological Program's Western Coniferous Forest Biome project. He and Dr. George Carroll, his collaborator at the University of Oregon, had undertaken the task of analyzing branch, foliage, and epiphytic communities of in the canopy of this old-growth forest. Their initial simplistic approach was based on the development of allometric equations, which were needed to estimate such variables, using traditional destructive methods: cut down the trees, sort them into their components, dry and weigh samples, and construct equations. However, this clearly was not going to work with giant trees that ranged up to eighty meters in height and two hundred centimeters in diameter at breast height! Nor did this approach adequately represent the three-dimensional structure of the canopy.

A solution to the dilemma was soon suggested and then proven by two graduate students, Diane Nielsen and Diane Tracy, who were also avid mountain climbers: utilize traditional climbing techniques and equipment to provide repeated nondestructive access to the canopies of these trees. Rigging of many trees followed along with development of booms or "spars" that provided access to lateral branches and branch-based sampling protocols.

Denison, Carroll, and the rest of their research team went on to develop their pioneering work in temperate canopies of old-growth forests using these techniques and without casualties. Early findings included descriptions of canopy architecture, foliar mass and area, and epiphytic communities in old-growth coniferous forests of the Pacific Northwest. Process-oriented research included the discovery of the important role of nitrogen-fixing canopy lichens, such as *Lobaria oregana*, in the nitrogen budget of these forests. At about the same time, but independently, similar canopy access methods were being pioneered in tropical forests by Meg Lowman in the Old World tropics and by Don Perry in the New World tropics (see Chapter 1).

This research could be viewed as the beginning of modern canopy studies, crude though it was by current standards. Equipment and techniques for canopy access, instrumentation, and sampling procedures—including the incredible capability existing today for collecting, managing, and analyzing large, spatially explicit data sets—have evolved dramatically in the forty years since that beginning.

Methods for accessing forest canopies have proliferated and continue to evolve. Rope-climbing techniques are highly sophisticated and routine, even in the largest of trees, such as the coast redwood (*Sequoia sempervirens*) and giant sequoia (*Sequoiadendron gigantea*). Cranes and walkways provide access to canopies for even the most handicapped of scientists, students, and citizens. Canopy cranes are notable for allowing the use of bulky or heavy instrumentation for in-situ canopy research that would otherwise not be possible. They have also demonstrated their value in allowing access to both the uppermost and outermost extremities of forest canopies and to fragile structures, such as standing dead trees or snags, which are otherwise inaccessible for in-situ research. Many are operational throughout the world, although none currently exist in North America.

Immense progress has occurred in research methodology and instrumentation—and this is what most of this book is about. Perhaps not fully appreciated by current generations of canopy researchers are the incredible advances in hardware and software that have made it possible to organize, store, access, and analyze the huge, spatially explicit data sets that modern instrumentation makes possible—and to do it on personal computers and share it via the web. When Denison and Carroll began their studies, we were still in the era of IBM punch cards and massive centralized computers, where the overnight model run most often yielded a single page reporting "formatting error"!

Our knowledge of canopy ecosystem structure, function, and composition has expanded exponentially over the last forty years, but we are only at the beginning of our discoveries. Many of those have been unexpected surprises that we sometimes refer to as epiphanies: soils and root systems in tree canopies; old trees that perpetuate crowns and foliage via epicormic branches and shoots; long-lived leaves; the incredible diversity of epiphytes, invertebrates, and even vertebrates rarely or never seen below the canopy; the amazing capacity for water transportation; and many others.

Even so, we are still only at the dawn of our understanding regarding the functioning and biodiversity of the forest canopy universe and their relationships to canopy architecture. An infinitude of discoveries—many unforeseen or even currently unimaginable—remain to be made. Some will involve generalities, so-called principles or theoretical constructs, and others will be specific to the immense diversity of forest types and conditions. All will require the laborious and dedicated efforts of field-oriented scientists, whose efforts this book is intended to serve.

PREFACE

This book is dedicated to the next generation of *arbornauts*—the technical term for those adventurous scientists who risk gravity, thorns, and stinging ants to explore the higher reaches of forest ecosystems. No work has focused on approaches and techniques for canopy study—although Andrew Mitchell did collate an overview handbook almost ten years ago (Mitchell et al. 2002). Since that time, the field has changed significantly and throughout the world, and forests have become a research priority, especially with the advent of climate change and the understanding of ecosystem services. Issues such as carbon sinks, biodiversity conservation, and canopy–atmosphere interactions have created a plethora of new methods and an urgency to measure the forest canopy accurately. As programs like the National Ecological Observatory Network (NEON; http://www .neoninc.org), the National Science Foundation–funded Long-Term Ecological Research (LTER) sites, and Earthwatch Institute's multimillion-dollar forest training program for HSBC Bank employees gain momentum and require increasingly labor-intensive methodologies, protocols for sampling become increasingly important to maximize time, funds, and energy.

This book updates the history of approaches to studying forest canopies and surveys the advances in techniques in view of the rapidly changing environmental priorities of the new millennium. Because different questions require a variety of approaches and techniques, this volume is not intended to standardize methods but rather to present advantages and disadvantages of selecting certain approaches and methods for particular objectives. The purpose of writing a book about canopy methods is threefold:

1. Share established methods and "recipes" for best practices of both observational and experimental techniques for studies of forest canopies.

2. Foster compatibility between different canopy researchers, different forests, and different projects, so that data collected can be compared both within and among research programs to invariably enhance cost-effectiveness, time, and energy.

3. Establish protocols that will lead to more effective forest conservation, using more rigorous data sets that are collected using best practices established by experts in the field.

The authors collectively represent more than one hundred years' experience in canopy research. Despite this lengthy service to the field, we sought additional experts for all aspects of methods and attempted to tap colleagues who deploy both old as well as new techniques of canopy sampling. The book is intended for a diverse audience, including not only practitioners in the field of forest science but also students of ecology or policy or conservation management, government agencies who wish to understand how field biologists collect and analyze data, citizen scientists, and armchair naturalists and explorers who seek to understand how scientists study and explore forests. Despite our best efforts, this book will not pose solutions to the myriad sampling challenges in forest canopies. For example, the notion of defining vertical transects in forests varies with respect to overall forest structure; in some cases, five meters represents the uppermost canopy, and in other situations, ten meters represents a heavily shaded understory. Similarly, insect sampling may prove exceedingly variable in terms of replication, relative numbers, and both temporal and spatial constraints. The abundance of Diptera may be low in montane forests with relatively cold weather conditions, but they can exceed several thousand per cubic meter in lowland moist tropical forests, making one homogenous sampling methodology difficult to standardize.

In 1982, a slim booklet called *Reaching the Rain Forest Roof* was published as the outcome of an early canopy symposium at the University of Leeds, England. Just over ten years later, the first canopy textbook, *Forest Canopies* (Lowman and Nadkarni 1994), was published, with more than 25 chapters, 40 contributing scientists, and 624 pages. During the 1980s, canopy exploration had burgeoned, but as the initial textbook illustrates, it was mainly focused on access and the safety of exploration into what E. O. Wilson affectionately called "the eighth continent." Perry (1986) and Lowman (1999) both wrote personal accounts of forest canopy exploration, the latter published in five languages, which opened up canopy research to students worldwide, including K–12 classrooms and citizen scientists, instead of a few privileged researchers in developed countries. Soon after, the second canopy textbook was published, also called *Forest Canopies* (Lowman and Rinker 2004), totaling 604 pages that focused on hypothesis-driven research. In the last decade, a host of personalized canopy accounts was written (e.g., Lowman et al. 2006; Moffett 1998; Nadkarni 2008; Novotny 2008; Preston 2008, etc.).

By providing their stories, these scientists gave citizens, students, and policy makers greater insights into how canopy scientists think and work. By beginning to inform not only scientists but also citizens and decision makers, canopy science has matured into a full-fledged scientific discipline, with its own specialized researchers, techniques, driving questions, and global challenges.

Methods in Forest Canopy Research is organized like a recipe book, where a user can turn to one section without reading the entire book. It does not, however, simply provide specific recipes for exactly how to sample every component of forest canopies: such methodologies are changing constantly and no one method is appropriate for every forest type or for each nuanced hypothesis. Instead, this book offers a basis for how forest canopy methods have developed, why specific methods were designed for different temporal and spatial attributes of forest canopies, and what methods are most appropriate for addressing particular objectives. In sharing the evolution of the last forty years of this arboreal exploration, it exposes the aspects of canopy research that do not yet have adequate methods to address. As soon as this book is printed, new methods will undoubtedly be reported in the literature; this underlies the dynamic and constantly changing elements of current forest canopy research. For this reason alone, we wanted to document the emergence of this new field of forest biology and chart its course for the first forty years of active and dedicated methodology. To this end, we devised a website (http://www.canopymethods.com) to track new techniques and to link to aspects of canopy research, including databases that will ensure improvements in forest data collection. We hope that this virtual element of our canopy methods publication will not only expand the user group of our canopy methods, findings, and hypotheses but also facilitate the necessary updates as the field changes and expands. Over the next forty years, new techniques to document biodiversity; forest processes, including carbon storage; and the growth/mortality/dynamics of the forest, including restoration and impacts of climate change, will take center stage as the next generation of forest biologists tackle the still relatively unknown canopy region.

The first chapter of this book introduces the history of canopy science and why this component of forest research was overlooked until the last few decades. The remaining chapters are structured into sections that reflect different types of canopy research: Chapter 2 measures the canopy structure of different forest types. Chapter 3 describes canopy access methods and how they are executed for specific canopy studies. Chapter 4 surveys biological units within a canopy. Chapter 5 measures the diversity of canopy habitats, biota, and processes. Chapter 6 reviews canopy–atmosphere interactions. Chapter 7 measures canopy–forest floor interactions. Chapter 8 assesses the application of canopy biology to conservation and education. Each chapter has a section of suggested reading and a summary section that highlights the future of canopy research and discusses the ways in which sampling methodologies fall short of adequately addressing important hypotheses. We hope that this book will not only create methods protocols that can be replicated throughout forest canopies around the world but also inspire newer and improved techniques that will ultimately lead to better decision making for forest conservation and management. Throughout the chapters, boxes written by experts highlight "canopy methods recipes" that are currently considered best practices and/or represent case studies that illustrate creative techniques.

We express enormous gratitude to Dr. Chuck Crumly, editor extraordinaire from University of California Press. Over the years, Chuck has been a major force of inspiration for canopy biology and had an enormous impact on disseminating the field around the world through his professionalism in the publishing world. We also convey our thanks to dozens of colleagues who came out of their treetop eeyries to comment and add new methods to this text. The indomitable spirit and sense of inquiry-based studies by arboreal explorers has given canopy science a unique sense of *spirit des corps* over the last three critical decades of its development. And last but not least, we thank all the educators and policy makers who have decided that ecosystem services, biodiversity, and forest canopies are worthy of understanding: your work as decision makers to educate the next generation is perhaps more critical than the technical science itself.

ACKNOWLEDGMENTS

The authors extend their arboreal thanks to the international community of canopy scientists who have contributed their sweat, creativity, and inspiration more than four decades of research and exploration. This book would not have seen the light of day without the vision of our editor, Chuck Crumly, who always encourages us to reach for the sky (or at least the treetops). Tim extends his thanks to the Louisiana Agricultural Experiment Station and colleagues associated with the Wind River Canopy Crane Research Facility, Smithsonian Tropical Research Institute, and H. J. Andrews Experimental Forest and Luquillo Experimental Forest Long-Term Ecological Research sites. In addition to those sites listed previously, Meg conveys her appreciation to the North Carolina Museum of Natural Sciences staff and volunteers, in particular mentor and director Betsy Bennett and her amazingly organized assistant Cindy Bogan, without whom this publication would not have ever been completed. She is also grateful to colleagues at North Carolina State University, the Aldo Leopold Leadership Program, ATREE (Ashoka Trust for Research on Ecology and the Environment in Bangalore, India, where much of this was written); Alemayehu Wassie Eshete and the clergy of the Christian Orthodox Church in Ethiopia; and all the other stakeholders in emerging countries who may find this compilation useful to inspire forest conservation from a canopy perspective. Special funds from the TREE (Tree Research, Education, and Exploration) Foundation, Sonia and Larry Ewald, the National Science Foundation, Fulbright, and National Geographic sustained specific components of this book over the years. And finally, Meg thanks her sons Eddie and James for their tree-climbing enthusiasm throughout a childhood of arboreal exploration, whose ancestors would be proud, despite the absence of a tail!

1

SETTING THE STAGE
Canopy Research Emerges as a Component of Forest Science

Indeed, over all the glory there will be a canopy. It will
serve as a pavilion, a shade by day from the heat, and
a refuge and a shelter from the storm and rain.

ISAIAH 4:5–6

Botany needs help from the tropics. Its big
plants will engender big thinking.

E. J. H. CORNER, CAMBRIDGE UNIVERSITY, 1939

THE IMPORTANCE OF FOREST CANOPIES

Our ancestors were tree dwellers. Throughout human history, people have taken to the trees as safe havens, sites of special spiritual connection, and cornucopias for food, medicine, materials, and productivity (reviewed in Lowman 1999; Nadkarni 2008). In many tropical forest regions, indigenous people rely on forests for their livelihoods. Increasingly on a global scale, people and governments are beginning to recognize the importance of ecosystem services provided by forests, which link directly to human health (Perrings et al. 2010). Such benefits include medicine, food, shade, building materials, gas exchange capabilities, energy production, carbon storage, genetic libraries, water cycles, and spiritual/cultural heritage. In many religions, the clergy are stewards of the forests (e.g., Bossart et al. 2006; Cardelús et al. 2012; Lowman 2010; Wassie-Eshete 2007). Children in many cultures climb trees for recreation and build tree houses as an important component of their creative and spiritual links to nature (Louv 2005). With their billions of green leaves that produce sugars from sunlight, the treetops are the engines that support life and the basis of food chains throughout the planet. In an evolutionary sense, humans descended from ancestors in the treetops. Recent findings about ancient hominoids in Ethiopia indicate that human ancestors inhabited forests (not savannahs as was previously thought; White et al. 2009). Anyone who pauses at the zoo to watch a monkey cavorting in tree branches is amused, inspired, and subconsciously reminded of some arboreal sensation that tugs on our evolutionary memory banks.

In Papua New Guinea, a tribe called the Korowai still lives in the treetops, erecting amazing aerial houses accessible by twig ladders. It is speculated that their unusual habit of community tree houses evolved as a mechanism to escape enemies on the forest floor and provide a healthy environment above the dank, dark understory. Tree houses remain a recreational vestige of children and adults alike that inspires links between humans, their ancestry, and the natural world. Many famous people have escaped to childhood tree houses: John Lennon (of the Beatles), Winston Churchill, the Roman Emperor Caligula, and Queen Victoria when she was a young princess (Nelson et al. 2000). Recent medical findings indicate that children who play outdoors and learn about nature have better health and well-being (Louv 2005; Lowman et al. 2009).

Why do the treetops hold such a spiritual and scientific importance for cultures throughout the world? And why have scientists only recently begun to explore these heights for scientific discovery, after decades of studying a mere fraction of the forest understory? Relatively few unknown frontiers of exploration still exist in the twenty-first century, but the treetops are still considered a "black box" in science (Lowman 1999). The other "black boxes" for exploration include the ocean floor and soil ecosystems.

Forest canopies reputably house more than 40 percent of the biodiversity of terrestrial ecosystems (reviewed in Lowman and Rinker 2004; Wilson 1992). The combination of sunlight, fruits, flowers, and year-round productivity of the foliage in tropical rain forests provides the ideal conditions for an enormous diversity of inhabitants. Thousands of species of trees and vines produce a veritable salad bar for millions of insects that are in turn eaten by myriad reptiles, amphibians, birds, and mammals, and those primary consumers are eaten by secondary consumers such as harpy eagles, jaguars, and other carnivores. Finally, the cycle of life is completed when soil decomposers break down and recycle all matter for uptake of component nutrients into the canopy (Frost and Hunter 2007).

In addition to classic food chains using energy from sunlight to cycle water and nutrients through leaves to herbivores to carnivores/omnivores to decomposers and back to plants, forests house extra niches for other unique forms of life. Bromeliad tanks, tree crotches, leaf surfaces, and epiphyte communities host extra layers of life in forest canopies. For example, bromeliad tanks house virtual swimming pools in the sky that are home to an entire microcosm of microorganisms. Mosquito larvae, nematodes, tarantulas, katydids, shovel-tailed lizards, and canopy mammals live in and/or drink from these aerial watering holes. Some poison dart frogs trek all the way from the forest floor into emergent trees to deposit their eggs in phytotelms. Other unique canopy niches include the crotches of trees that provide germination sites for strangler figs and soil repositories that house many microarthropods usually associated with the forest floor. Strangler figs are the only trees known to start life "at the top" and send their aerial roots extending downward, eventually penetrating the soil below to expand and strangle their unwitting host plants. Epiphytes add an extra layer of biodiversity and productivity in the moist, sun-flecked branches. Even more amazing, the surfaces of canopy leaves provide a substrate for epiphylly, another extra layer of plant forms including lichens, mosses, and fungi, many of which grow exclusively on leaf surfaces. Within the "canopies" of these tiny epiphylls lives an entire microcosm of microinvertebrates and other microscopic organisms. Nothing rivals the forest canopy in terms of biodiversity—layers upon layers of life, all nurtured by sunlight, moisture, and warmth in a unique combination that fosters an extraordinary diversity and abundance of species.

This world of canopy plants, insects, birds, mammals, and their interactions remained relatively unknown and out of reach to scientists until as recently as twenty-five years ago (reviewed in Lowman 2004). William Beebe first introduced the world to the wonders of tropical rain forest canopies with his popular book *High Jungle* (1949), in which he described using a bow and arrow or climbers to hoist rope ladders into the high canopy. Field ecologists used slingshots to propel ropes into the treetops in the late 1970s, but at that time no one recognized that this green umbrella overhead was a critical component of global health. Exploration of both the deep sea and outer space was relatively commonplace prior to the exploration of forest canopies. But with the recent escalation of climate change and population pressures, canopies have become a proverbial "canary in the coal mine," since their declining health is a harbinger of environmental changes on a global scale, as well as a driver of further change (Foley et al. 2003; Janssen et al. 2008). Currently, forest canopy scientists—along with reef ecologists, ice physicists, soil biologists, water chemists, and others—are taking on the role of planetary physicians, working against a near-impossible timeline in hopes of unraveling the critical mysteries of how our planet functions. With access to forest canopies, our knowledge of the machinery of forest ecosystems has greatly expanded. And perhaps less appreciated in a technical sense, forest canopies enhance our sense of wonder and appreciation of the natural world, serving as drivers for more ambitious conservation agendas.

WHY SUCH A TIME LAG IN STUDYING FOREST CANOPIES?

Biologists in the nineteenth and twentieth centuries traditionally based their ideas about forests on observations made at ground level. These ground-based perceptions are summarized in a comment by Alfred R. Wallace in 1878: "Overhead, at a height, perhaps, of a hundred feet, is an almost unbroken canopy of foliage formed by the meeting together of these great trees and their interlacing branches; and this canopy is usually so dense that but an indistinct glimmer of the sky is to be seen, and even the intense tropical sunlight only penetrates to the ground subdued and broken up into scattered fragments . . . it is a world in which man seems an intruder, and where he feels overwhelmed" (Wallace 1878).

Binoculars and telescopes were the first documented tools for canopy exploration. Charles Darwin, in the nineteenth century, looked into the tropical rain forest foliage, exclaiming,

> Delight itself . . . is a weak term to express the feelings of a naturalist who, for the first time, has wandered by himself in a Brazilian forest. The elegance of the grasses, the novelty of the parasitical plants, the beauty of the flowers, the glossy green of the foliage, but above all the general luxuriance of the vegetation, filled me with admiration. A most paradoxical mixture of sound and silence pervades the shady parts of the wood. The noise from the insects is so loud, that it may be heard even in a vessel anchored several hundred yards from the shore; yet within the recesses of the forests a universal silence appears to reign. To a person fond of natural history, such a day as this brings with it a deeper pleasure than he can ever hope to experience again. (Darwin 1883)

Ideas about forest canopies changed very little for almost a hundred years from Darwin's day until the 1950s, when a steel tower was constructed in Mpanga Forest Reserve in Uganda to study gradients from the forest floor to the canopy. Towers provided access to monitor insect vectors of human diseases, representing one of the first applied biological studies conducted in the forest canopy (Haddow et al. 1961). Other early canopy studies were mainly exploratory by nature: Operation Drake by Andrew Mitchell through the Oxford University explorers club, the first canopy walkway structure in Malaysia anchored among tree crowns (Muul and Liat 1970); ladders affixed to observe tree phenology and animal visitors (McClure 1966); Kamal Bawa's early chromosomal cytology of canopy trees in the Himalayas (Mehra and Bawa 1968; Mehra and Bawa 1969); and Steve Sutton's early exploration of Sulawasi (Sutton 2001). All these early canopy pioneers utilized methods that were relatively inexpensive and designed as creative solutions to overcome the challenges of gravity but were not necessarily devised to be easily replicated for experimental design purposes or for multiple users.

The late 1970s represented a golden age of canopy access, with development in 1978 of single rope techniques (SRT), independently designed for ecological data collection by Meg Lowman (1983) in Australia and Don Perry (1986) in Costa Rica. Whereas scuba equipment in the 1950s heralded the age of exploration for coral reefs (reviewed in Sale 2002), ropes and harnesses inspired the "race to the top" (of trees). This versatile toolkit of ropes, harness, and climbing hardware enabled scientists to reach the midcanopy with ease, suspended from a rope to observe pollinators, epiphytes, herbivores, birds, monkeys, and even sloths. Portable and relatively inexpensive, SRT and double rope techniques (DRT) allowed even budget-limited graduate students to survey life in the treetops (reviewed in Mitchell et al. 2002). Ropes were ineffective, however, to reach the leafy perimeters of tree crowns, since the ropes had to be looped over sturdy branches usually close to the tree trunk. To access the uppermost foliage of canopy trees, new devices were invented to overcome earlier limitations. For example, botanists in Indonesia devised the canopy boom, a horizontal bar with a bosun's chair at one end, which could be swung around the leafy canopy away from the woody trunks. In Pasoh, Malaysia, a combination of ladders, ropes, and booms launched research that solved the mystery of the pollination of dipterocarp flowers (Appanah and Chan 1981). Bawa (1969) used scaffolds and ropes to study pollinators in tropical trees in India. In temperate forests with their lower canopies, Lowman adapted construction scaffolding to survey the leafing phenology of birch trees in Scotland (Lowman 1978).

Engineers and creative canopy biologists partnered to construct canopy bridges and platforms in the 1980s. The first two canopy walkways were constructed nearly simultaneously: one in Malaysia, anchored in tree crowns by Ilaar Muul, and another in Queensland, Australia, supported by telephone poles (see Lowman 2009). Built in Lamington National Park, the Australian walkway was a collaboration of the ideas of Lowman and Peter O'Reilly, owner of the ecotourist lodge called O'Reilly's Guesthouse. Lowman's Earthwatch programs in Queensland rain forests required a canopy access tool that held groups of researchers (since ropes are solo operations). More than five years after the walkway construction in Australia and Malaysia, North America's first canopy walkway was built in 1992, using a suspension-bridge construction design suspended between oak trees in Massachusetts (Lowman and Bouricius 1995), and America's first public canopy walkway was constructed among Florida oak-palm hammocks in Myakka River State Park in Florida in 2000 (Lowman et al. 2006). Canopy walkways have since been replicated throughout the world, using a modular construction design developed by Canopy Construction Associates (http://www.canopyaccess.com), Greenheart, and a few other companies. Throughout the last decade, canopy walkways and ladders used in conjunction with climbing ropes, zip lines, and other tools have become popular ecotourism destinations as well as research tools. (See also http://www.treefoundation .org.) Subsequently, canopy walkways can provide sustainable income to local people and are especially useful in tropical rain forests because they provide an economy aside from logging (Lowman 2009).

Perhaps one of the most creative canopy access tools is the French-designed hot-air balloon, called *Radeau des Cimes* (trans. "raft on the rooftop of the world"). The balloon flies independently but also operates in conjunction with an inflatable raft (27 meters in diameter) that can be set on top of the canopy surface to serve as a base camp or platform atop the uppermost branches of tall trees (Hallé and Blanc 1990). In 1994, the *Radeau des Cimes* expedition team pioneered a new technique in French Guinea called the sled, or skimmer. This small (5 meters across), equilateral, triangular miniraft was towed across the canopy by the dirigible, similar to a boat with a trawling apparatus in the water column. The sled allowed rapid exploration between trees to compare pollinators, photosynthesis, herbivores, and the relative diversity/abundance of canopy life.

Construction cranes represent the most recent tool for safe access into the forest canopy (reviewed in Mitchell et al. 2002). In 1990, the Smithsonian Tropical Research Institute first erected a 40-meter-long crane in a Panamanian seasonally dry forest; since then, ten other crane operations have commenced operation in Australia, Switzerland, Germany, Japan, Indonesia, the United States, and Venezuela. Cranes are expensive to install and operate (usually ranging from $1 million to $5 million) but offer unparalleled, repeated access to the uppermost canopy within reach of the crane arm.

CANOPY DIVERSITY

The forest canopy is defined as "the top layer of a forest or wooded ecosystem consisting of overlapping leaves and branches of trees, shrubs, or both" (Parker 1995). Tree canopies represent the hotspots of the forest—new leaves, flowers, pollinators, birds, arboreal mammals, orchids, lizards, mosses, and millions of insects. The bulk of energy captured from sunlight is concentrated in this region high above the forest floor. Oxygen is just one of the byproducts of this canopy engine, which represents the interface of Earth and atmosphere.

Studies of plant canopies typically include four organizational levels of approach: individual organs (leaves, stems, or branches), the whole plant, the entire stand, and the forest landscape. Canopy biology represents a relatively new discipline of forest science only formally launched over the past three decades (albeit linked inextricably to whole-forest biology). Canopy science incorporates the study of mobile and sessile treetop organisms and the processes that link them to the larger ecological forest ecosystem. Studying mobile bird populations is dramatically different in methods, tools, and temporal dynamics from studying sessile lichens anchored to leaf or bark surfaces. With the recent urgency surrounding climate-change research, forest canopies have emerged as an important interface between Earth and atmosphere; episodes such as insect outbreaks or canopy species extinctions serve as early warning signals indicative of hotter, drier climates.

Access to forest canopies has led to the discovery of millions of species inhabiting this above-ground world. Quite by chance, Dr. Terry Erwin of the Smithsonian Institution first discovered the abundance of life in the treetops. Erwin sprayed several tree canopies in the tropics with a mild insecticide, and an astounding diversity of arthropod residents fell to the ground in a heap, enabling him to count and catalogue them (Erwin 1982). From this initial harvest of insects in Panamanian rain forest trees, Erwin extrapolated that there may be 30 million species on our planet, not 1–2 million, as previously estimated. Since that time, studies by other canopy biologists in Australia, Indonesia, and Peru confirmed Erwin's predictions that millions of insects inhabit forest canopies.

Field biologists who focus on biodiversity seek to catalog, identify, and understand the role of all creatures on Earth. This is not simply a naming game; its ultimate purpose is to understand the structure and function of an ecosystem, almost the same way that we seek to know how the components of a car engine operate together to create an efficient machine. The challenge to discover and identify species throughout the world is not easy. Finding a new beetle in the treetops is akin to locating the proverbial needle in the haystack—90 percent perspiration and 10 percent luck. All organisms collectively—orchids, beetles, birds, vines, frogs, and many others—constitute biodiversity, otherwise known as the variety of species on Earth. The word *biodiversity* has become politically and scientifically important over the past two decades, as human activities have accelerated ecosystem degradation and subsequent loss of species throughout the world.

Long before Erwin, naturalist Charles Darwin in the 1800s estimated that approximately 800,000 species inhabited the Earth. (One can only imagine that the Queen of England was most impressed by his scientific prowess in calculating this apparently enormous number!) Nearly a hundred years later, Erwin's canopy-fogging data raised Darwin's original tally more than forty-fold. As a result, scientists now believe that the treetops are home to a large proportion of biodiversity on the planet. Only the soil microcosm is predicted to exceed canopy biodiversity, but biologists have not yet learned how to accurately measure organisms underfoot. Professor Edward O. Wilson, eminent biologist at Harvard University, speculated that as many as 100 million species may inhabit our Earth based on initial surveys of the canopy, soil, and oceanic biodiversity combined (May 2010).

Many canopy species provide essential ecosystem services on which human beings depend for survival. In the Amazon, plants produce chemical defenses against insect attack; these chemicals, in turn, are used by indigenous cultures for medicinal purposes. The shaman (or medicine man) is a highly respected community leader who has inherited generations of knowledge about the practice of using plants for medicinal purposes. Canopy leaves, barks, and fruits provide a veritable apothecary in the sky, all of which have evolved over time due to the unique interactions of plants with their herbivores (mostly insects). Canopies provide not only medicine but also food, industrial compounds, and construction materials, and they serve as reservoirs providing for soil conservation, genetic libraries, carbon storage, and freshwater conservation. Additional ecosystem services (a concept that links to economic values) include gas exchange, energy production, climate control, and cultural/spiritual sanctuaries for many people (reviewed in Lowman 2009; Schowalter 2011; White et al. 2010).

FUTURE DIRECTIONS IN CANOPY SCIENCE

The tools and discoveries of canopy biology burgeoned in the 1980s, and subsequent quantitative and experimental fieldwork expanded in the 1990s. Analogous to the three-dimensionality and complexity of coral-reef research (Sale 2002) but building on the rigorous experimental design of two-dimensional intertidal ecology (Murray et al. 2006), canopy biology has experienced a dramatic evolution in its short history, from pure exploration and a sense of wonder to increasingly quantitative approaches, embracing a tremendous urgency to protect forests and their critical canopy functions (not just data collection) before they disappear.

Over these past three decades, the role of canopy biologists has changed. No longer can scientists dangle leisurely from the trees and contemplate the beauty of harpy eagles and katydids; instead, they are caught up in an urgent race against time to provide answers to critical questions about canopy effects on climate before the potential effects of deforestation become a reality. To date, only 1.7 million of an estimated 30 (or perhaps 10? 100?) million species have been identified. At current rates, taxonomists classify approximately 7,000 new species per year, an agonizingly slow page compared to the threats of deforestation. The effort of sorting, counting, and naming as many as 100 million species is daunting. The ecological task of determining which species are most important to forest health is even more challenging. As Stewart Udall, former US Secretary of the Interior, once said, "Over the long haul of life on this planet, it is the ecologists, and not the bookkeepers of business, who are the ultimate accountants."

How many is 100 million species? Is there a way to make that enormous number more meaningful for non-mathematicians? If 200 scientists discovered and identified one new species every day for the rest of their lives, they would need nearly 1,500 years, including weekends and holidays, to complete their task of identifying the estimated biodiversity on Earth (Lowman et al. 2006). Even more urgent than names alone, biologists need to determine benchmarks for evaluating changes in forest canopy conditions and effects on Earth's environment. Is biodiversity important? And do we care if some species become extinct? How much forest and which tree species are critical to maintain this global machinery that we call a forest ecosystem? Can forests function in fragments, and will they remain healthy if replanted in single-species plantations? Unfortunately, no one has answers to these important questions. Biologists have not studied forest canopies long enough to understand the processes that are critical to their health. In the *Sand County Almanac* (1949), world-renowned ecologist Aldo Leopold wisely said, "To save every cog and wheel is the first precaution of intelligent tinkering." Another famous scientist, Paul Ehrlich, considered biodiversity analogous to the mechanical parts of an airplane. He speculated that, if an airplane mechanic continued to remove nuts and bolts from a plane, the machine will eventually cease to fly. Similarly, as species disappear, we may find that our ecosystems can no longer "fly"—that is, support life (Head and Heinzman 1990; Meadows 1987). The critical question asked most often still remains: Will such extinctions reach a critical threshold beyond which humans cannot exist?

More than sixty years later, Leopold's words still hold true. We need to preserve all the pieces of ecosystems (i.e., species) until we identify which ones are essential to the operation of the machinery. Forest canopies represent "home" to a disproportionately large number of species. Because forests are such efficient machines, producing energy, medicine, materials, fibers, food, nutrient cycling, and atmospheric gases critical to all life on Earth, the continued health of forests is directly intertwined with human health.

Since researchers first observed canopy pollinators in the 1960s and the first canopy walkways were built more than twenty years ago, millions of hectares of tropical rain forest have disappeared, along with thousands of as-yet-undiscovered new species. Loss of forest canopies is a critical issue for our children and their children as they grow up to inherit the stewardship of our planet. Earth is full of exciting discoveries relating to forest canopies: new medicines; exotic perfumes; ecosystem services such as fresh water supply and carbon sequestration; keystone predators and other food-chain dynamics; and important economic products such as chocolate, corn, oranges, coffee, and rubber, to name but a few. The next decade is critical. Forest canopies are essential to healthy ecosystems that translate into sound economic policies.

This book is organized to facilitate selection of appropriate methods for particular canopy research objectives. Chapter 2 describes methods appropriate for comparing canopy variables among forest types. Chapter 3 describes methods for accessing forest canopies within particular sites. Chapter 4 surveys methods for describing the structure of tree crowns and forest canopies. Chapter 5 describes methods for describing habitat diversity and measuring the abundances and distribution of canopy organisms, as well as canopy processes. Chapter 6 reviews methods for measuring canopy–atmosphere interactions, and Chapter 7 describes methods for measuring canopy–forest floor interactions. Finally, Chapter 8 assesses the application of canopy biology to canopy conservation and education programs. Throughout the book, our objective has been to provide an overview of available methods and their advantages and disadvantages for particular research objectives. We do not intend to recommend particular methods, given that selection of methods must be consistent with particular research objectives. However, we encourage collaboration among canopy researchers and selection of common methods that will facilitate comparison of canopy variables among regions and forest types in order to advance canopy science globally.

FOREST TYPES AND SITE CHARACTERISTICS

Forests are ecosystems characterized by dominance of long-lived woody plant species. Forests originally covered an estimated 50 percent of the Earth's ice-free land surface, but anthropogenic deforestation has reduced this area significantly. The world's forest cover has been cleared to about half of its historic extent (Matthews and Hammond 1999; Myers 1999; Perry et al. 2008; FAO 2001; Waring and Running 2007; reviewed September 2010 in http://www.economist.com/specialreports). Despite losses, humans continue to depend on forests for a variety of ecosystem services, including wood for building material and fuel, food from plants and associated animals, fresh water, cultural and recreational uses, and carbon sequestration (Millenium Ecosystem Assessment 2005; Schowalter 2011). Describing forest variables is critical in evaluating the value of these services and justifying conservation efforts.

Forest ecosystems typically have complex three-dimensional structure that provides habitat and food resources for a high diversity of other organisms. In addition, forests have considerable capacity to modify local and regional environmental conditions, through shading of the ground, stabilization of soil, storage of water and nutrients in woody biomass, and moderation of turbulence and evapotranspiration that drive local and regional precipitation. Much of this capacity depends on the interaction between forest type and site characteristics that determines forest structure (see Chapter 4), which in turn determines canopy habitat conditions, biodiversity and ecological processes (Chapter 5), and interactions with the atmosphere (Chapter 6) and forest floor (Chapter 7).

Forest types show distinct structures (Fig. 2.1A–D), largely as a result of climate and disturbance history, and vary in their productivity, biodiversity, and contributions to human well-being. Forest type can be categorized broadly as latitudinal zones of tropical, temperate, or boreal forest (and as evergreen versus deciduous) and more finely as subtypes dominated by particular tree species. Site characteristics include topography, climate, soil type, and disturbance regime. The species dominating forests in different biogeographic regions vary in their phylogenetic histories; for example, *Quercus* spp. dominate temperate forests in the northern hemisphere, whereas *Nothofagus* spp. dominate temperate forests in the southern hemisphere. However, considerable convergence is apparent among forests with similar site characteristics.

FIGURE 2.1.

Different forest types, including: A) Temperate forest showing foliage in autumn just prior to leaf fall in Vermont; B) Conifer forest at Wind River crane site, Oregon; C) Cool temperate or montane cloud forest in Queensland, Australia; and D) monodominant palm forests in subtropical Florida.

Tropical forests occupy the equatorial belt between the Tropic of Cancer and the Tropic of Capricorn. Consistently warm temperatures and high precipitation in the equatorial convergence zone provide a long, often year-round growing season that contributes to high primary production and rapid growth of trees and associated organisms. Because solar exposure is direct and intense, sufficient light is available to support trees of various sizes from understory to canopy emergents. However, competition for growing space, water, and nutrients is intense, leading to considerable specialization to exploit available resources. Consequently, diversity of tree and other species is very high.

Temperate forests occupy latitudes between 25° and 45° N and S of the equator (Fig. 2.1A–B). These zones show marked seasonality in temperature because of the tilt in Earth's axis as it circles the Sun, with cold winters that interrupt the growing season. Temperate deciduous forests are distinct because trees lose their leaves annually, and many canopies turn brilliant colors when the chlorophyll dies after initial frost events (Fig. 2.1A). Trees dominating temperate forests typically are adapted to survive freezing temperatures, including insulating bark and retrieval of nutrients from foliage prior to dropping their leaves in the fall. Some temperate tree species, particularly coniferous species in northern latitudes (Fig. 2.1B), are evergreen, withstanding freezing temperatures and exploiting the freedom from competition by deciduous species for light and water during fall and early spring to maintain relatively high productivity throughout the year (Waring and Franklin 1979). Pines also tend to be relatively tolerant to drought and fire and often dominate temperate forests on frequently disturbed sites.

Boreal forests dominate latitudes above 50° N and S. These zones have long cold seasons adverse to plant growth and are dominated by coniferous species, especially of *Abies* and *Picea* in northern regions, that are tolerant of extreme cold and capable of sufficient photosynthesis during the short growing season to survive long periods of dormancy. These trees typically are evergreen to exploit the short period that is suitable for photosynthesis, because growth cannot occur until soils rise at least a few degrees above freezing (Waring and Running 2007).

SITE CHARACTERISTICS

Among the most important characteristics affecting the development of particular forest types are climate, topography, soil type, and disturbance regime. These factors interact to produce complex landscape gradients of forest patches distinguished by tree species composition, height, density, and three-dimensional canopy structure.

Climate varies widely within latitudinal zones, as a result of global circulation patterns and land form (see below). Precipitation largely dictates the distribution of particular forest subtypes. For example, most tree species in tropical rain forests are evergreen, but seasonal variation in precipitation results in dry seasons of varying duration and intensity that may restrict the survival of species unable to tolerate these conditions. Some tree species in forests with longer dry seasons are drought deciduous and produce new foliage with the return of wet conditions. In extremely dry tropical regions, short thorn-scrub forests predominate. On the other hand, persistent cloudiness at high elevation limits solar exposure, leading to less diverse cloud forests dominated by trees and other species that can tolerate lower light conditions.

Similarly, beech (*Fagus americanum*) and hemlock (*Tsuga canadensis*) dominate mesic temperate forests in eastern North America, oak (*Quercus* spp.) and hickory (*Carya* spp.) dominate intermediate forests, and pines (*Pinus* spp.) dominate more arid sites. Although tropical forests are reputably diverse, many stands are not (Fig. 2.1C–D). For example, the cool temperate or montane rain forests of Australia are predominately *Nothofagus* (Fig. 2.1C), and many subtropical and tropical forests feature palms (Fig. 2.1D).

Topography interacts with global circulation patterns to influence regional temperature, precipitation, and soil moisture. Mountains interrupt airflow and direct moist air upward, where moisture condenses and falls as precipitation on the windward side, supporting growth of relatively dense and tall forests. Dried air flowing down the leeward side increases evaporation of water, leading to relatively arid conditions and development of shorter, sparser woodlands and thorn-scrub forests. Higher elevations are cooler and moister than are lower elevations (Fig. 2.1C). Temperature and moisture conditions also vary among aspects of elevated land. Hillsides and mountainsides that face east or north typically are cooler and moister than those facing west or south.

Lower elevations and coastal areas with poor drainage, where water stands on the surface for long periods of time, support only those tree species that can withstand waterlogging and root hypoxia (e.g., cypress-tupelo and mangrove swamps). Although swamp forests are associated with standing water, many tree species require seasonal drying for seed germination in unflooded soil to maintain adequate recruitment.

Soil or substrate type constitutes another filter on what tree species grow at a site. Fine et al. (2004) used reciprocal transplants of clay- and white sand–specialist forest plant species and herbivore exclosures in a lowland Amazonian site in Peru to demonstrate that clay specialists grew significantly faster than white sand specialists on both soil types when protected from herbivores. However, when unprotected from herbivores, clay specialists dominated clay forests and white sand specialists dominated white sand forests, demonstrating that associated herbivores also affect plant distribution. Serpentine soils and lava flows are other examples of soil or substrate types that restrict tree species to those capable of tolerating these conditions.

A number of site conditions contribute to the formation of canopy gaps—empty spaces in the canopy that form as a result of abrasion between crowns that create "canopy shyness," unequal growth rates among trees, treefall, and disturbances (Brokaw 1982). Gaps range in size from the spaces between individual branches or tree crowns (Connell et al. 1997; Putz et al. 1984) to spaces that occupy the entire vertical profile to the forest floor (Bongers 2001; Brokaw 1982). Gaps between crowns provide flyways for birds, and larger canopy openings provide necessary light and space for species intolerant of shade or competition. Gaps developing as forests age greatly increase the complexity of canopy surface topography and interaction with the atmosphere (see Chapter 6). Depending on the size of canopy gaps and environmental conditions, gaps may remain open, be colonized and filled by shade-intolerant tree species, or be closed from the side by branch elongation from surrounding trees.

Disturbances are relatively abrupt events in time and space that substantially alter habitat and growing conditions across landscapes (Schowalter 2012; Walker and Willig 1999; White and Pickett 1985; Willig and Walker 1999). Disturbances vary in type (e.g., fire, storm, drought), magnitude (physical forcing and severity of effect on biota), frequency (number of events per time period), and extent (size of area affected). Storms and fire are acute disturbances that affect individual trees in time spans measured in hours, whereas flooding and drought can affect trees over much longer periods of time. The effect of a disturbance event depends on the unique combination of type, magnitude, time since the last event, and scale. Forests that experience particular disturbances frequently, relative to the life span of dominant species, tend to evolve adaptations to minimize vulnerability or injury (e.g., buttress roots to increase wind firmness or flame-resistant or insulating bark to minimize fire damage). Forest canopies can buffer the effects of disturbances to a large extent, depending on canopy structure and intensity and scale of disturbance. However, extreme changes in conditions created by severe disturbances result in local extinction of susceptible species and elevated populations of others that can exploit postdisturbance resources or freedom from competitors, herbivores, or predators.

Forests in all parts of the globe are subjected to disturbances (Fig. 2.2). Major disturbances include drought, fire, flooding, cyclonic and other storms in tropical forests (Heartsill-Scalley et al. 2007; Van Bael et al. 2004; Whigham et al. 1999), ice storms and windstorms in temperate forests (Binkley 1999), and ice and snow storms in boreal forests (Engelmark 1999; Taylor and MacLean 2009). In some forests, disturbances interact with biotic responses to increase the likelihood of future disturbance (Jenkins et al. 2008; McCullough et al. 1998; Taylor and MacLean 2009).

Disturbance events affect forest canopies and associated organisms differentially. For example, a low-intensity ground fire may scorch the lower canopy but only burn individual trees near accumulated fuels, whereas a catastrophic crown fire may burn and kill most trees and their associated canopy organisms over large areas. Frequent cyclonic storms can preclude emergence of overstory tree crowns in tropical forests and cause widespread breakage or treefall (Fig. 2.3). Drought leads to moisture stress in plants, resulting in premature leaf fall and, often, outbreaks of insect herbivores (Mattson and Haack 1987; Van Bael et al. 2004).

Disturbances affect community composition by locally exterminating intolerant species and redistributing organisms and resources (Schowalter 2012). For example, the survival of canopy organisms during a fire depends on their tolerance range to extreme temperatures and to the temperatures reached in various parts of the canopy.

FIGURE 2.2.

Examples of forest disturbances: (A) fire, (B) hurricane, (C) landslide, and (D) volcanic eruption.
From Schowalter (2012) with permission from Taylor and Francis.

High winds and landslides can move organisms from their original locations. Torres (1988) reported that Atlantic hurricanes transported a number of species from North Africa to Caribbean forests. Flooding in Amazonian forests transports seeds and organisms to new sites where probability of survival may or may not be improved.

Disturbances also alter postdisturbance environmental conditions in ways that favor species that may require the new habitats and/or resources, thereby maintaining higher biodiversity on a landscape scale. New colonists tend to be generalists that are capable of surviving under a wide range of environmental conditions in the absence of intense competition. Such species initiate a process of recovery known as ecological succession. Succession represents a gradient of forest subtypes through time, similar to gradients that may exist among patches of different successional stages on the landscape (Salo et al. 1986; Van Cleve and Viereck 1981). Successional stages vary in their canopy structure, including surface topography and function, and in their contributions to primary production, energy and matter fluxes, and interactions with atmospheric and forest-floor conditions.

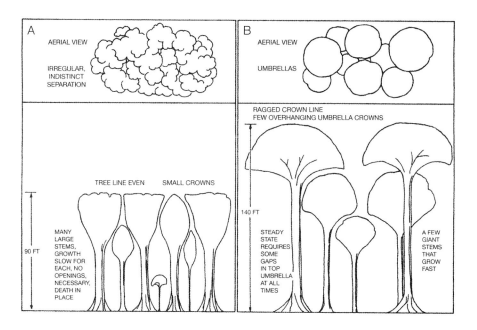

FIGURE 2.3.
Differences in tropical forest structure resulting from hurricanes: (A) a hurricane-dominated forest with even canopy surface maintained by breakage of emerging crown and (B) a canopy structure with emergent crowns typical of tropical forests infrequently disturbed by storms.
From Odum (1970).

Increasingly adverse growing conditions near the range boundaries of forest types limit tree density and height growth. Broad ecotones often occupy boundaries between forest types and between forests and other ecosystems, blurring the geographic pattern of forest types. For example, moisture limitation and frequent fires along forest borders with grasslands tend to maintain relatively open-canopied woodland ecotones; severe winter storms and ice packed at high latitudes or elevations result in ecotones of sparse, stunted trees along the timberline with tundra ecosystems. Such ecotones may shift geographically through time as environmental conditions change (Allen and Breshears 1998; Gear and Huntley 1991). Describing and distinguishing these forest types requires statistical discrimination.

METHODS FOR DESCRIBING FOREST TYPE AND SITE CHARACTERISTICS

Selection of appropriate forest type or subtype is crucial for a number of research objectives. Forests vary widely in their structure and primary production (Chapter 4), suitability as a habitat for various organisms (Chapter 5), interactions with atmospheric chemistry and air flow (Chapter 6), and maintenance of water flux and soil fertility (Chapter 7), as well as in their contributions to various ecosystem services (e.g., timber, water yield, fish and wildlife production, recreational uses, carbon sequestration). Conservation and restoration of key structures, processes, and services requires comparison and evaluation of their values in various forest types.

Different forest types require different approaches for research on canopy structure, diversity, or effects on global processes. For example, data may be collected relatively easily from the ground in short or sparse forests, whereas tall forests require labor-intensive climbing techniques or expensive towers or cranes for data collection in the canopy. Difference in canopy structure could enhance or limit some remote-sensing techniques. Although the three-dimensional complexity of many forests would seem to defy description by remote-sensing techniques, remote-sensing technology is developing rapidly, and sensors are now able to measure foliage height profiles through complex multilayered canopies (reviewed in Nadkarni et al. 2004). In fact, remote sensing may provide the best method for measuring canopy conditions over large areas that are inaccessible from the ground.

GROUND BASED

Traditional methods for measuring areas of various forest types required access from the ground and sight lines through the canopy for measurement of tree heights, crown separation or gap size, and so on. This method is labor intensive and hazardous in dense, multicanopied forests—especially tropical forests. However, such methods are still useful for research in some forests (Birnbaum 2001) and may be necessary to confirm remotely sensed data.

REMOTE SENSING

Remote sensing has become an extremely valuable tool for measuring the area and distribution of forest types and subtypes (Box 2.1). Differences in multispectral reflectance among tree species and laser-guided detection of height profiles provide for rapid optical imaging analysis, using moderate resolution imaging spectroradiometry (MODIS) and light detection and ranging (LiDAR) to measure global canopy height and species composition (e.g., Lefsky 2010), making the labor-intensive and often inaccurate ground-survey methods of the past unnecessary, except for confirmation. Alternatively, a potentiometer can be connected to the lens of a single-lens reflex camera through a gear system and, using the focusing feature, calibrated for various distances to measure canopy height (Birnbaum 2001). Many subcanopy structural features that distinguish forest types also are amenable to measurement by remote sensing. For example, foliage height profiles help to distinguish forest types and successional stages.

Data on canopy surface topography, crown form, height, and so on can be analyzed using multivariate statistical techniques to distinguish forest types. Digitized images from remote-sensing photographs can then be used to measure areas of each forest type, subtype, or age class, regionally and globally.

BOX 2.1: LOW-TECH AND HIGH-TECH APPROACHES TO REMOTE SENSING OF FOREST CANOPIES

Stephanie Bohlman

Remote sensing has become an important method for studying forest canopies because it gives a perspective on the forest that is not easily obtained from ground-based studies. Remote sensing views the forest from above and on landscape to global scales, whereas field studies are biased toward the understory and small spatial scales. In the past, remote-sensing images were often too difficult to obtain, too hard to interpret, and too costly for the average ecologist. But with the availability of digital photography, increased accessibility of high-resolution satellite images, and widespread familiarity with geographic information systems (GIS) and remote-sensing software, remote sensing is becoming a routine tool in a canopy biologist's toolbox. While some remote-sensing devices are incredibly sophisticated and need complex algorithms for analysis, others are very accessible, budget friendly, and easy to interpret. I discuss some of these low-tech and high-tech approaches and their contributions to forest canopy biology.

Aerial photography has been providing information about forest canopies for many decades and is one of the oldest forest canopy "methods." Aerial photography has been used to quantify the volume of timber but also gives information about the shapes and conditions of individual tree crowns (e.g., Aldrich and Drooz 1967). Digital photography and digital videography have made the capture and analysis of aerial photography more accessible and useful to ecologists. For example, to map the distribution of three species of canopy palms on Barro Colorado Island (BCI), Panama, Jansen et al. (2008) took a mosaic of photographs with a digital camera from a small airplane over the whole island (14 km²). GPS readings allowed us to pinpoint the locations of each photo, and GIS software and a high-resolution satellite image (Quickbird) allowed us rotate and warp each image to obtain ground coordinates from the photos and an image mosaic for the whole island. From the high-resolution photo mosaic (0.1 m² resolution), we mapped canopy trees of three species of palms for the whole island and used this map for estimating canopy fruit abundance (Jansen et al. 2008).

More sophisticated remote-sensing methods, such as light detection and ranging (LiDAR) and stereophotographs, can give structural information on individual tree canopies. If the images can be precisely georeferenced, remote-sensing data and field data taken on the same trees can be synthesized. As an example, we used stereophotographs, taken from a helicopter by a professional photogrammetrist, to map every canopy tree crown over 11 hectares in the 50-hectare forest dynamics plot on BCI. By viewing the forest in three dimensions using specialized computer equipment and software, we measured tree height and precisely mapped and digitized crown boundaries. While we did this by hand, it is now possible to use segmentation software to automate these mapping procedures (Palace et al. 2008). In the field, we used the image-derived crown map to link the mapped crowns to the stems in the plot census. This has allowed us to characterize, for the first time, the species composition of the canopy for a tropical forest stand. With this map, we have compared the species diversity of the canopy versus the whole forest, quantified the large distances that canopy trees can displace their crowns in order to access direct sunlight, determined that the species identity of a canopy tree affects the growth of trees growing directly underneath its crown, and used the information as direct input to forest dynamics modeling (Bohlman and Pacala 2012).

(continued)

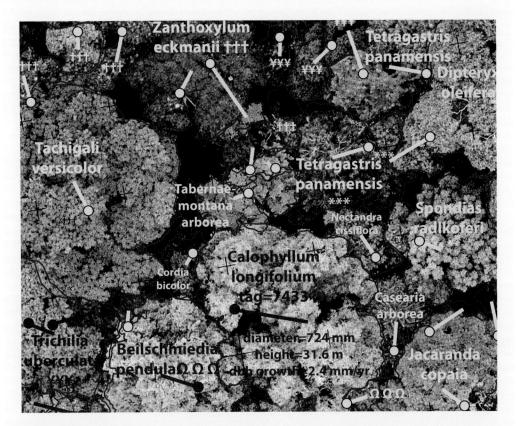

BOX 2.1, FIGURE 1. Bohlman's scaling from tree to forest as part of new imaging processes. High-resolution stereophotographs of Barro Colorado Island, Panama. Canopy tree crowns were digitized to provide heights and sun-exposed crown area. The crowns were matched to tags in the forest inventory census to provide species identity (shown on map) and twenty-year diameter history. Lines extend from the trunk (circles) to the crown center of each crown. Species with more than one crown are shown with symbols (ΩΩΩ for *Beilschmiedia pendula*, *** for *Tetragastris panamensis*, ¥¥¥ for *Trichilia tuberculata*, ††† for *Zanthoxylum eckmanii*).

Photo © Rich Grotefendt, Tropical Forest Management Foundation, grotefen@uw.edu.

(continued)

Some of the most promising remote-sensing techniques for studying trees use high-tech, high-cost, multisensor airborne sensor platforms (Anderson et al. 2008; Asner et al. 2007; Brown et al. 2005). Platforms that combine different sensors can give detailed information about the canopy, although these methods are still under development. Multiple-return LiDAR gives information on canopy structure (Asner et al. 2008; Treuhaft et al. 2009). Hyperspectral images give species-detection capability (Clark et al. 2005) and information on chemical composition (Asner et al. 2005). True color images provide visual information needed for interpreting the other types of information and tying the image data to ground data. Access to hyperspectral, LiDAR, and multisensor images remains limited, as only a few academic, government, and commercial organizations have such systems, but hopefully they will become more widely available over time.

GEOGRAPHIC INFORMATION SYSTEMS (GIS)

Mapped data can be overlaid on a grid of square cells (raster method), with each cell defined by its x- and y-coordinates becoming a computerized data point (see Waring and Running 2007). Grid cells can be any size, so any landscape can be defined within the database to any degree of accuracy. Alternatively, polygons of any shape can be used to define a landscape (vector method). Cell identification within a vector database is more difficult than in a raster format, but a vector format represents complicated landscapes more realistically.

GISs provide a mechanism for overlaying map layers depicting different canopy and site data. For example, a digitized map of canopy height superimposed on maps of climate variables, soil variables, disturbance history, and surface topography can be used to evaluate the effects of these site variables on canopy height or to evaluate the combined effect on global forest productivity, carbon flux, water yield from forested watersheds, or other ecosystem services (Waring and Running 2007; see Chapter 7). Geostatistical tools (e.g., kriging) permit interpolation between sample points to fill in areas with missing data. A number of software packages are available for these analyses.

MODELING

Computerized modeling has become a powerful tool for simulating hypothesis testing and predicting changes in canopy structure and interactions with atmospheres or forest floors over time. Several models of forest development are available, including the BIOME-BGC and CENTURY models (e.g., Parton et al. 1993; Waring and Running 2007), which have been used to predict effects of climate change on distribution of forest types, effects of changes in forest composition or canopy cover on global processes (Waring and Running 2007), or effects of atmospheric nitrogen deposition and herbivory on forest carbon and nitrogen dynamics (Throop et al. 2004). Clearly, the experimental manipulation of forests over large areas (with adequate replication) to test hypotheses concerning global carbon sequestration or climate change is impossible (see Chapter 3). However, data from a number of experiments on a smaller scale around the globe can be entered into a sufficiently accurate model to simulate a global experiment. Furthermore, data of adequate quality for all the required model variables are necessary to ensure robust model output (Table 2.1). Methods for measuring values for such models are described primarily in Chapters 4, 6, and 7. Chapter 5 describes the effects of associated canopy organisms on these processes.

SUMMARY

Forests are defined by the dominance of large woody trees and occupy a substantial portion of the globe, even with substantial anthropogenic deforestation and conversion. Distinct forest types can be distinguished across latitudinal gradients, with tropical forests centered on the equator, temperate forests between 25° and 45° N and S latitudes, and boreal forests above 50° N and S to timberline ecotones with tundra. Within these broad forest types, subtypes distinguished by canopy height, canopy layering, and phylogeny of dominant taxa reflect variation in climate, topography, and disturbance history. Forests recovering from disturbances typically are composed of distinct species assemblages adapted to greater light and resource availability.

Forest types can be distinguished by their canopy surface topography and height profiles, with tropical forests showing multilayered canopies with dominant trees emerging above the general canopy, having umbrella-shaped crowns to intercept solar radiation from directly above; temperate forests showing less layering and distinct seasonality in foliage cover; and boreal forests showing tall columnar crowns to intercept solar radiation primarily from the side. Gaps of various sizes develop as forests age, as a result of unequal tree growth, tree fall, and disturbances. Canopy surface topography of older forests becomes very complex, compared to the more uniform canopy height and smooth surface topography of younger, especially managed, forests.

The measurement and depiction of forest types has advanced rapidly with the advent of remote sensing, GIS, and computerized modeling techniques. Remote sensing allows for the imaging of forests on a global scale, and GIS techniques facilitate mapping and measurement of areas covered by different forest types. The overlaying of map layers representing different databases permits the evaluation of correlations among multiple variables. Modeling allows for hypothesis testing and the prediction of effects of changes in forest cover and global climate, given that the experimental manipulation of forests on a large scale is impossible or, at least, impractical.

Measures of area covered by forest type facilitate the selection of forest sites for more detailed research, as described in the following chapters, based on their representation of forest conditions.

TABLE 2.1. Data Requirements for the FOREST-BGC Model, with
Examples of Values for Coniferous and Broadleaf Forests*

Parameter and Units	Conifer Forest	Broadleaf Forest
Leaf C, kg ha^{-1}	2,400	2,400
Soil C, kg ha^{-1}	300,000	300,000
Litter C, kg ha^{-1}	100,000	100,000
Soil N, kg ha^{-1}	30,000	30,000
Litter N, kg ha^{-1}	2,000	2,000
Leaf turnover, year^{-1}	**0.25**	**1.00**
Stem turnover, year^{-1}	0.01	0.01
Coarse-root turnover, year^{-1}	0.01	0.01
Fine-root turnover, year^{-1}	0.5	0.5
Leaf lignin fraction, kg C lignin kg^{-1} C	**0.25**	**0.125**
Fine-root lignin fraction, kg C lignin kg^{-1} C	0.25	0.25
Specific leaf area, ha kg^{-1} C	**0.0025**	**0.0045**
Canopy light extinction coefficient	−0.5	−0.5
Precipitation interception coefficient, m H$_2$O projected LAI^{-1} day^{-1}	0.0002	0.0002
Snowmelt temperature coefficient, m H$_2$O °C^{-1}	0.005	0.005
Snow albedo decay coefficient, °C^{-1}	0.004	0.004
Snowpack energy deficit, °C	5.0	5.0
Spring minimum leaf water potential, mPa	0.5	0.5
Leaf water potential at stomatal closure, mPa	1.65	1.65
Maximum stomatal conductance, m H$_2$O s^{-1}	0.006	0.006
Cuticular conductance, m H$_2$O s^{-1}	0.00005	0.00005
Leaf boundary-layer conductance, m H$_2$O s^{-1}	0.001	0.001
Canopy aerodynamic conductance, m s^{-1}	0.2	0.2
Optimum stomatal conductance temperature, °C	20.0	20.0
Maximum stomatal conductance temperature, °C	40.0	40.0
Vapor pressure deficit, Pa		
Start of conductance reduction	750.0	750.0
Completion of conductance	3,000.0	3,000.0
Leaf maintenance respiration coefficient, day^{-1}	0.0012	0.0012
Stem maintenance respiration coefficient, day^{-1}	0.0001	0.0001
Coarse-root maintenance respiration coefficient, day^{-1}	0.0001	0.0001
Fine-root maintenance respiration coefficient, day^{-1}	0.0012	0.0012

TABLE 2.1. (*continued*)

Parameter and Units	Conifer Forest	Broadleaf Forest
Q_{10} for plant maintenance respiration	2.0	2.0
Ratio of all-sided to projected LAI	2.3	2.0
Ratio of sapwood, kg C m^{-2} LAI^{-1}	**0.25**	**0.35**
Leaf translocation fraction	0.5	0.5
Q_{10} for soil maintenance respiration	2.4	2.4
Maximum leaf / (leaf + root) ratio for allocation	**0.70**	**0.80**
Maximum coarse root/stem ratio for allocation	0.25	0.25
Critical litter C:N ratio	25.0	25.0
Critical soil C:N ratio	10.0	10.0
Fine-root N fraction, percent $N_{fine\ root}$ percent N_{leaf}^{-1}	0.5	0.5
Stem and coarse-root N fraction, percent N_{stem} or percent $N_{c\text{-}root}$ percent N_{leaf}^{-1}	0.01	0.01
Leaf growth respiration fraction, kg^{-1} C	0.35	0.35
Stem/coarse-root/storage growth respiration fraction, kg^{-1} C	0.30	0.30
Fine-root growth respiration fraction, kg^{-1} C	0.35	0.35
Maximum storage fraction	0.20	0.20
Living fraction of sapwood	**0.08**	**0.10**
Water stress integral factor	0.15	−0.15
Fraction N_{leaf} in Rubisco	0.10	0.10
Maximum soil decomposition rate, year^{-1}	0.03	0.03
Maximum N uptake by roots (V_{nmax}), kg N kg^{-1} $C_{fine\ roots}$	0.05	0.05
N uptake Michaelis-Menten constant (K_n), kg N ha^{-1}	20.0	20.0
Maximum average N_{leaf} concentration, kg N kg^{-1} C	0.04	0.04
Minimum average N_{leaf} concentration, kg N kg^{-1} C	0.02	0.02
Leaf off, year day	**0.0**	**140.0**
Leaf on, year day	**365.0**	**260.0**
Fine root on, year day	60.0	60.0
Fine root off, year day	304.0	304.0
Volumetric saturated soil water content	0.44	0.44
b parameter for soil water potential	−6.09	−6.09
Soil matric potential, Mpa	0.002224	0.002224

* For very generalized simulations, only the variables in bold type may need to be specifically defined; default values can be used. When more site- and/or species-specific data are available, the model can incorporate these additional details

From Waring and Running (2007) with permission from Elsevier.

SUGGESTED READING

Kuuluvainen, T. 1992. Tree architectures adapted to efficient light utilization: Is there a basis for latitudinal gradients? *Oikos* 65: 275–84.

Lefsky, M. A. 2010. A global forest canopy height map from the moderate resolution imaging spectroradiometer and the geoscience laser altimeter system. *Geophysical Research Letters* 37, L15401. doi:10.1029/2010GL043622.

Perry, D.A., R. Oren and S.C. Hart 2008. *Forest Ecosystems,* 2nd Ed. Johns Hopkins University Press, Baltimore, MD.

Throop, H. L., E. A. Holland, W. J. Parton, D. S. Ojima, and C. A. Keough. 2004. Effects of nitrogen deposition and insect herbivory on patterns of ecosystem-level carbon and nitrogen dynamics: Results from the CENTURY model. *Global Change Biology* 10: 1092–105.

Waring, R. H., and S. W. Running. 2007. *Forest ecosystems: Analysis at multiple scales.* 3rd edition. San Diego, CA: Academic Press.

3

CANOPY ACCESS METHODS
*Making It Possible to Study the Upper
Reaches of Forests Accurately and Safely*

The first explorers and scientists who accessed tree canopies found a new world of discovery and inspiration. E. O. Wilson called it "the last frontier" of biological research on the planet (Wilson 1992). Andrew Mitchell referred to its invisible inhabitants as "a world I could only dream of" (Mitchell 2001). Tom Lovejoy confessed that "the canopy rendered me the biologist's equivalent of Tantalus from the very outside" (Lovejoy 1995). Steve Sutton compared it to "Alice grows up," as canopy science moves from a sense of wonder to a reality of hypotheses (Sutton 2001). Nalini Nadkarni exclaimed about "tree climbing for grown-ups" (Nadkarni 1995), and Meg Lowman simply commented, "My career is not conventional. I climb trees." (Lowman 1999). In 1985, these six individuals may have represented almost half of the canopy scientists worldwide. Today, just over two decades later, several hundred researchers focus almost exclusively on Wilson's "last frontier."

This chapter summarizes the diverse spectrum of canopy methods—including a short history of their development and the evolution of advancements in the safety of access tools and significant twentieth-century methods that served as drivers to expand the field of forest science by opening up canopy research frontiers (from camera traps to construction cranes, parataxonomists to international collaborative expeditions, sensory devices on dragon flies to satellite imagery, and hardcover field guides to applications downloaded into handheld mobile devices). Of critical importance is that the array of current canopy access techniques facilitates a broad range of research, and one method is usually preferred over another to address specific hypotheses and monitoring programs. The chapter ends with speculation about the missing links still on the drawing boards that will facilitate future exploration of forest canopies and seek urgent solutions to maintenance of sustainable forest resources. For example, can canopy scientists develop reliable methods to estimate canopy cover with respect to forest health and degradation? Can carbon sequestration be accurately mapped in different forest types? Why are mammals and other biodiversity overlooked and/or inaccurately assessed in estimates of deforestation? Will insect outbreaks exacerbated by climate change impact forest health? How can biodiversity in the upper canopy be accurately surveyed and ultimately managed for conservation of their genetics libraries? Will the influx of smart phones and handheld applications allow citizen scientists to collect additional data points that will assist with forest conservation and management?

THE DILEMMA OF MOBILE VERSUS SESSILE ORGANISMS: DESIGNING METHODS TO ACCURATELY MEASURE DIFFERENT CANOPY COMPONENTS

Forest canopy is defined as "the top layer of a forest or wooded ecosystem consisting of overlapping leaves and branches of trees, shrubs, or both" (Art 1993; Parker 1995). Studies of plant canopies typically include four organizational levels of approach: individual organs (leaves, stems, or branches), the whole plant, the entire stand, or the plant community (Lowman 1984; Lowman 1995; Ross 1981). Canopy biology is a relatively new discipline of forest science that incorporates the study of mobile and sessile forest organisms and the processes that link them as an ecological system (Fig. 3.1). The development of canopy research has been affected by several spatial and temporal constraints of this habitat (reviewed in Lowman and Moffett 1993):

1. Differential use of this geometric space by canopy organisms
2. Variation in the microclimate of the canopy–atmosphere interface
3. Differences in ages within the canopy (e.g., soil/plant communities growing in crotches or leaves of different ages between sun and shade environments)
4. Heterogeneity of surfaces (e.g., bark, air, foliage, debris, and water)
5. High diversity of organisms, many of which are unknown to science

Recognizing these criteria, biologists need to design useful access techniques to test hypotheses in the canopy (see Chapter 5). Canopy studies range from measuring sessile

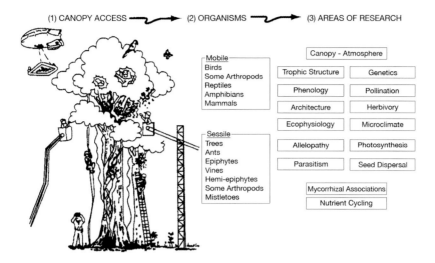

(1) CANOPY ACCESS → (2) ORGANISMS → (3) AREAS OF RESEARCH

Mobile
Birds
Some Arthropods
Reptiles
Amphibians
Mammals

Sessile
Trees
Ants
Epiphytes
Vines
Hemi-epiphytes
Some Arthropods
Mistletoes

Canopy - Atmosphere
Trophic Structure | Genetics
Phenology | Pollination
Architecture | Herbivory
Ecophysiology | Microclimate
Allelopathy | Photosynthesis
Parasitism | Seed Dispersal
Mycorrhizal Associations
Nutrient Cycling

FIGURE 3.1.
Forest canopy research has progressed from the passive canopy observations at ground level to the design of methods for studying different types of organisms and more rigorous experimental approaches for documenting canopy interactions and processes.

organisms (orchids, sedentary insects, trees) to mobile organisms (flying insects, birds, mammals) to canopy processes (studies of the interactions of organisms up in the tree-tops). All these studies require access methods that are effective at variable heights and different spatial and temporal scales and can function in an air substrate, oftentimes while dangling from a rope or in some otherwise awkward position (awkward for the researcher going against gravity, that is). How are organisms detected and sampled in such a hetero-geneous environment where humans are rendered less agile? In a scenario similar to the expansion of coral reef fish ecology in the 1960s with the advent of scuba (Sale 2002), can-opy biologists in the 1980s and 1990s developed access protocols to measure the spatial, temporal, and substrate heterogeneity of their environment (Lowman and Moffett 1993).

STUDIES OF SESSILE ORGANISMS IN FOREST CANOPIES

These pose fewer logistical challenges to measurement than other canopy components because they can more easily be counted without escaping. Sessile organisms include the obvious groups such as trees and vines, as well as more cryptic organisms (e.g., mosses, lichens, scale insects). Trees are the largest sessile organisms in forest canopies, and they comprise the major substrate of the canopy ecosystem. Tree species—with their varying architecture, limb strength, surface chemistry, and texture—play fundamental roles in shaping the canopy community. Epiphytes (or air plants) live on the surfaces of trees, as do many epiphylls (tiny air plants, such as lichens that live on leaf surfaces). In addition, trees serve as both shelter and food for many mobile organisms, and most canopy processes are directly dependent upon trees. The biggest obstacle to quantitative studies of canopies is access to the growing tips (buds, flowers, root hairs) that usually exist in the uppermost branches where the light is greatest. Some methods, such as the canopy raft and the crane, provide access to this fragile region, whereas others such as ropes and walkways are limited to midcanopy regions since they need to be near sturdy branches for safety. Major achievements in studies of sessile organisms include the extensive documentation of canopy architecture (e.g., Hallé et al. 1978); the distribution and biology of epiphytes (Benzing 1990; Colwell et al. 2008; Nadkarni 1984); mapping canopy surfaces from construction cranes (Parker et al. 1992); and the growth of vines (Putz and Mooney 1991).

BOX 3.1: CONSTRUCTION OF CANOPY WALKWAYS

Philip Wittman

Research in the canopies of forests (Lowman and Wittman 1996) has been limited by the logistical constraints of safe and easy access (reviewed in Mitchell 1982; Moffett and Lowman 1995). Of the many methods used to gain access to the heights, only a few allow several people to be in close proximity or in the canopy for long periods of time. Observation platforms and canopy bridges are used to overcome these limitations, but the design and costs of such structures often exceed research budgets (Box 3.1, Fig. 1). Compared to suspension bridges, the canopy bridges discussed here are lighter and easier to construct and install and can be placed higher in the canopy. Walkways can also be integral to forest conservation, by creating local income streams through ecotourism (Lowman 2009a).

WALKWAY SYSTEMS

A canopy walkway system typically incorporates some combination of platforms, bridges, and means of access (Lowman and Bouricius 1995). A canopy bridge as discussed here consists of an overhead safety cable, two handrail cables, and a treadway. The treadway consists of two strong and flexible supporting cables with slender traversing treads, separated by a small spacer, distributed along their entire lengths. The ends of each cable are attached to large trees that are stabilized with guy cables. The treadway hangs freely and assumes a special shape called a *catenary*.

BOX 3.1, FIGURE 1. The world's longest canopy walkway allows research and ecotourism activities in Amazon rain forest canopies near Iquitos, Peru.

DESIGN CONSIDERATIONS

Prior to the design phase, an on-site survey visit must be made so appropriate trees can be selected and accurate measurements can be made in and between the trees. This typically requires two experienced climbers with at least one ground-support person and can take two to four full days, depending on the number of trees selected and their location, height, architecture, and so on. Trees are climbed to measure the girth at platform height, locate support branches (distance, angle, and size), and measure the distance between trees in the canopy. Additionally, local sources of materials can be investigated during this time as well.

To design and construct a walkway, one must know the attributes of all the materials used, such as the tensile strength of the cables and the weights of all component parts. For our treadways, we prescribe safety factors of five or higher (WRTB 1993). To achieve this safety factor, the tension on the cables must be either measured or calculated. To permit easy negotiation, the slope of the treadway should not be too large (i.e., less than twenty-five degrees for the angle from the horizontal at any place on our treadways). The first choice for the supporting cable is 3/8 inch stainless steel aircraft cable; however, calculations may reveal that the required safety factor cannot be met or that the angles at the ends of the treadway are too great. In this case, a higher tensile strength (thicker) support cable must

(continued)

(continued)

be used. Different choices for other hardware, including eyebolts, guy cables, or even trees, might be necessary. Once the final specifications are determined, a design can be drafted and approved. Then necessary materials are purchased and transported to the site. Once all materials are on site, the construction crew can begin the construction of the canopy access system without delay.

CONSTRUCTION CONSIDERATIONS

To specify a particular treadway catenary requires two parameters, one of which is the horizontal distance between the two ends, defined as the *span* (Bouricius et al. 2002). The most feasible way to ensure during construction that a treadway meets its design specifications is to measure the sag. A line-of-sight method using a portable laser level works well. The sag can be changed during construction by varying the length of the support cables. This is accomplished by adjusting the cables through the cable clamps and eye bolts at one end of the treadway.

The means of entering the canopy access system is driven by cost and the type of user. Towers with stairs are the most user friendly but also the most costly. Other means (in order of decreasing expense but increasing effort) of getting to the canopy include climbing ladders (bolted to the tree trunk), being pulled up in a block and tackle bosun's chair, or even climbing a rope.

COST ESTIMATES

The on-site survey cost ranges from $2,000 to $6,000 plus travel expenses, which are highly variable depending on the country and the location within the country.

Due to the custom nature of each canopy access system, in scope, access and location costs are highly fluid. The most expensive scenario is a client who simply wants to have a complete system built without any interaction on their part. A cost estimate then must include all travel, room and board, labor, customs fees, transportation costs, and materials costs. Estimates are $300/foot of completed treadway; platforms range from $2,000 to $10,000, depending on size and complexity; and towers with stairs are roughly $10,000 per every 10 feet in height. These costs are negotiable with input and help from the client, who is often able to provide room and board, local laborers, and appropriate locally available materials (especially the lumber).

MOBILE ORGANISMS IN THE CANOPY

These pose great challenges for scientists because studying mobility in a complex three-dimensional space is difficult. The enormous wealth of biodiversity in forest canopies was recognized only after access by single-rope techniques (SRTs) allowed for observations above the understory (Lowman 1984; Perry 1986). Subsequently, the development of canopy-fogging methods facilitated research to quantify millions of arthropods inhabiting the treetops (Erwin 1982). Most studies of vertebrates were historically made from ground level, including extensive work on the biology and distribution of neotropical mammals (e.g., Emmons 1995). In contrast, Jay Malcolm (1995; 2004) adapted the indigeneous *peconha* climbing technique to trap small rodents in the tree canopies of Brazil and made the discovery that many have unexpected arboreal proclivities. He found that species exhibited distinct height preferences and that more mammals were arboreal than terrestrial. Similarly, Peter Taylor was one of the first biologists to census small mammals in temperate deciduous oak-beech canopies; he found that one species, the flying squirrel, conventionally censused from ground or understory levels, was an important predator on the gypsy moth (Taylor and Lowman 1996). These examples illustrate how improved access to the canopy has corrected misinformation about the distribution and abundance of forest organisms and forest processes. Similarly, an emerging literature on an entire country's canopy fauna is based on canopy access techniques, serving as an important driver for conservation and effective forest management (e.g., Ganesh and Devy 2006).

Ornithologists also face the challenge of trying to observe and/or capture birds in tree crowns. In New Guinea, Bruce Beehler hoisted nets up and down tall poles to quantify birds of paradise in the canopy (Beehler 1991). Lovejoy and Bierregaard (1990) found that birds increased the size of their territories vertically to compensate for forest fragmentation. In Peru, Charlie Munn used a large slingshot to position aerial mist nets in the canopy (1995), and Nalini Nadkarni and Terry Matelson (1991) used SRTs to observe 193 species of birds using epiphytes in the cloud forests of Costa Rica.

Reptile and amphibian populations in tree canopies were studied in Puerto Rico. Reagan (1995) used single-rope access to monitor *Anolis* lizard populations in tree canopies. Roman Dial and Joan Roughgarden (1995) also used single-rope access to perform one of the first experimental canopy manipulations by excluding lizards from tree crowns and monitoring changes in insect populations. With the advent of serious global declines in amphibians, population studies of frogs in both canopy and forest-floor locations have accelerated (e.g., Houlahan et al. 2000; Laurance et al. 2010; Lips 1998). Invertebrate studies in forest canopies have perhaps received more attention and subsequently endured more controversy than any other component of aerial ecology.

Terry Erwin's canopy fogging techniques to count invertebrates in tropical tree crowns have led him to speculate that 32.5 million species exist on Earth, 2.5 million higher than his previous calculations (Erwin 1982; Erwin 1983; Erwin 1990; Erwin 1991). Novotny's data from Papua New Guinea suggest that 8–10 million species might be more accurate (Novotny et al. 2006, but see May 2010). Nonetheless, the enormous spatial and temporal variability of insects in tree crowns, as well as artifacts of sampling, make studies of canopy arthropods difficult, with subsequent enormous volumes of data to analyze (reviewed in Basset 1997; May 2010; Table 3.1).

TABLE 3.1. Estimate of the Number of Species on Earth through the Biological Literature

Publication	Number of Species
Darwin 1865	800,000
Metcalf 1940	1 million
Southwood 1961	10 million
Erwin 1982	30 million
Stork 1988	10 million
Gaston 1991	5–10 million
Erwin 1992	32.5 million
Wilson 1992	100 million
Novotny 2006	8–10 million
May 2010	8–10 million

BOX 3.2: IBISCA: A COLLABORATIVE PROGRAM TO STUDY THE DIVERSITY AND DISTRIBUTION OF ARTHROPODS FROM CANOPY TO FOREST FLOOR (HTTP://WWW.IBISCA.NET)

Maurice Leponce, Bruno Corbara, and Yves Basset

IBISCA (Investigating the Biodiversity of Soil and Canopy Arthropods) is an international research program that aims to study the spatial (horizontal, vertical, and altitudinal) and temporal distribution of the organisms that constitute a major part of forest biodiversity: arthropods. Interactions with plants (mostly trees) and selected other organisms are also studied in this context.

The IBISCA approach is based on highly integrative research projects and the use of state-of-the art canopy access techniques. Field projects are performed worldwide in tropical, subtropical, and temperate forests. So far, projects have been implemented in Panama (2003–4); Queensland, Australia (2006–8); Esperitu Santo, Vanuatu (2006); and Auvergne, France (2008–9; Basset et al. 2007; Corbara 2009a, b).

IBISCA was initially designed by Yves Basset (Smithsonian Tropical Research Institute, Panama), Bruno Corbara (Université Blaise Pascal, Clermont-Ferrand), and Héctor Barrios (University of Panama) in response to the lack of large datasets on the diversity and distribution of arthropods at multiple scales (horizontal, vertical, and temporal) in tropical rain forests. Such data are fundamental to understanding the structure of arthropod communities from the ground to the canopy and to testing hypotheses about the origin and maintenance of this biodiversity.

IBISCA-Panama and the subsequent IBISCA projects are articulated around various

BOX 3.2, FIGURE 1. Standardized 20 × 20 meter IBISCA plot where multiple collection methods are used at multiple heights: (A) insecticide fogging, (B) Malaise trap, (C) flight intercept and butterfly traps, (D) sticky traps, and (E) beating. Multiple focal taxa include representatives of (F,G) pollinators (e.g., bees, butterflies), (H) herbivores (e.g., caterpillars), (I) sap suckers (e.g., Hemiptera, Nogodinidae), (J) fungus feeders (e.g., Coleoptera, Erotylidae), (K) predators (e.g., ants), and (L) decomposers (e.g., termites). Photos by Y. B. (A), J. S. (B, J), and M. L. (C–I, K–L).

central scientific questions—namely, about quantifying the horizontal and vertical species turnover in a lowland tropical rain forest (IBISCA-Panama), quantifying altitudinal species turnover and its implication in the context of climatic changes in a continental (IBISCA-Queensland, led by Roger Kitching, Griffith University) or insular (IBISCA-Santo led by Corbara) location, or quantifying species turnover among deciduous forest types in a temperate region (IBISCA-Auvergne led by Corbara).

The IBISCA protocol is as follows:

(continued)

(continued)

1. Multiscale, allowing a comparison of the diversity and abundance of organisms studied among sites, strata (ground, understory, canopy), and seasons. This is of particular importance, since the local species richness (alpha-diversity) and the species turnover among sites (horizontal beta-diversity) may differ between the ground and the canopy. Consequently, the species turnover from the ground to the canopy (vertical beta-diversity) may differ from local to regional scale (Basset et al. 2007). Without a three-dimensional approach, it is difficult to assess whether the diversity is highest at ground or canopy level. At landscape level, a stratified sampling is conducted, including the relevant habitats, strata, and microhabitats (Leponce et al. 2010). In addition, temporal replicates are performed whenever possible since the distribution and abundance of organisms may vary seasonally (e.g., according to their life stage). This seasonality affects the catchability and the ease of identification of the specimens: immature arthropods are often very difficult to identify. Temporal replicates yield a representative picture of the functioning and dynamics of arthropod assemblages.

2. Multitaxic, since no single arthropod taxa is representative of the pattern of distribution and abundance of the others, each requiring its own ecological requirements (Lawton et al. 1998). Because an all-taxa approach would have been extraordinarily costly in terms of time and resources, a functional approach including representative taxa of various life-history strategies (e.g., bees as pollinators, termites as scavengers, hemiptera as sap suckers) was adopted. Depending on the project, nonarthropod organisms interacting with arthropods are also included (e.g., host plants or vertebrate predators).

3. Multimethod, since the appropriate collection method may differ according to focal taxa and because the use of numerous collection methods compensate, to some extent, for the inevitable bias of each individual method in terms of representativeness of the real abundance of each taxa in the community (Leponce et al. 2010). In addition, it may help developing new and efficient protocols for studying particular arthropod assemblages. A special emphasis is given to methods that allow a direct comparison among forest strata such as sticky traps, flight-intercept traps, Malaise traps, and light traps (Basset et al. 2007).

A characteristic of the IBISCA modus operandi is that all projects are integrated and that all data are included in a collective database. This integration concerns data collection, processing, and analyses.

During field work, participants concentrate their sampling in the same set of plots. A typical IBISCA plot measures 20 × 20 meters (400 m² is also the area of the SolVin-Bretzel canopy raft). Within the plot, a botanical team identifies all trees that are also tagged. Entomological teams follow botanists in increasing order of disruptive protocols (e.g., insecticide fogging comes last). Tree climbers assist scientists to set up the collection devices in the canopy. All participants use preprinted labels with unique sample codes that are the only valid codes for the collective database. Depending on the location, participants are based in various infrastructures already present or built for the occasion: a proper laboratory (Panama), national park buildings (Queensland), field camps (Santo), or even a castle under restoration (Auvergne). Whenever possible, samples, especially those obtained from massive collection methods, are preprocessed on site. This is executed with the help of assistants supervised by professional taxonomists involved in the fieldwork. These assistants are either students, volunteers, or skilled amateur naturalists. Parataxonomists (Basset et al. 2004) are also ideally suited to collect the primary biodiversity information (e.g., collect samples and environmental data, rear specimens, and conduct field experiments) and process the material collected (e.g., sort, prepare, preidentify, image, and add specimens to the database) but have not been hired yet. A central processing of the material is ideal to be able to track the specimens from collection to identification. A collective database allows for the linking of specimens to samples and for performing global ecological analyses. For each taxonomic group, a workgroup leader is designated. He is the contact person toward the project coordinators and database managers. He coordinates the work of expert taxonomists and is responsible for the homogeneity of (e.g., in terms of systematics) and feedback on data. Besides taxonomic workgroups, other workgroups are dedicated to special techniques (e.g., canopy access) or data analyses (e.g., multivariate statistics).

In terms of canopy access, the assistance of professional tree climbers is a constant in all IBISCA projects. IBISCA-Panama succeeded in gathering a large array of additional tools (Box 3.2, Fig. 2): a canopy crane, the SolVin-Bretzel canopy raft, the canopy bubble (a helium balloon), and a tree observatory (the Ikos). The prototype of a new hot air/helium balloon, the canopy glider, was tested in Santo and Auvergne.

IBISCA is an international project in essence, with more than a hundred participants (field scientists, assistants, and taxonomists) from up to fifteen countries. Core participants are selected according to both their competence and their team spirit, which is a key ingredient of the program. The choice of focal taxa is also matched with the availability of committed and enthusiastic investigators. Funding of IBISCA projects is supported by both public (i.e., regional governments, universities, museums, national parks, and other national institutions) and private partners (i.e., private companies or foundations). In addition, many participating scientists independently obtain support from their national granting bodies. Except in the case of the Queensland project, IBISCA projects were managed by the nongovernmental organization Pro-Natura International, which previously organized several expeditions with the canopy raft.

(continued)

BOX 3.2, FIGURE 2. Canopy access methods used during IBISCA projects (photomontage, not to scale). Clockwise, from bottom left: Ikos treehouse, canopy bubble, SolVin-Bretzel canopy raft, canopy crane, and canopy glider.
Photos: M. L., Y. B., and N. Baben. Photomontage: I. B.

(continued)

Journalists, photographers, documentary filmmakers, and artists also participate in the expeditions, to improve public awareness of biodiversity studies.

IBISCA has set a new "industry standard" involving team work (taxonomists, ecologists, and students) and international collaboration and complimentary skills, from the field to the laboratory. The IBISCA program has completed four major biodiversity field surveys in three biogeographical regions and convened three scientific workshops. Baseline data were obtained on key topics, such as the spatio-temporal distribution of arthropods and vascular plants. For example, the Panama database contains approximately 100,000 records documenting the spatio-temporal distribution of half a million specimens in a lowland tropical rain forest. Methodological advances have been achieved by testing a large array of collection methods. For example, it appears that flight interception or light traps installed in the canopy help obtain a more complete image of termite assemblage than traditional ground-based methods (Bourguignon et al. 2009). The same method also helps collect a surprisingly wide range of taxa, rare or absent in other protocols, such as Issidae (Gnezdilov et al. 2010). In terms of innovative canopy access techniques, the canopy glider has demonstrated its maneuverability at the upper canopy level. The transfer of techniques designed for a complex tropical environment to a temperate one helped capture for the first time a global image of the distribution of arthropods from the ground to the canopy in temperate forests. Baseline data from IBISCA-Queensland are used to identify which taxa are responding with greatest sensitivity to the climatic changes currently associated with the different altitudes and to design predictive models and monitoring tools to inform management decisions in a changing climate. The experience gained through IBISCA also helped design the protocol of the Arthropod Monitoring Initiative of the Center for Tropical Forest Science (http://www.ctfs.si.edu/group/arthropod monitoring). In addition, numerous new species were discovered and are in the process of being formally described and named. This will take a few years, depending on focal taxa. IBISCA has fostered international collaborations, with scientists from fifteen countries involved in the projects. Without such intensive collaboration, a comprehensive understanding of the functioning of complex forest ecosystems is not possible. Future IBISCA projects will further develop innovative ways to sample canopy organisms and to cost-efficiently process a large amount of biodiversity samples, which remain major challenges when conducting large-scale biotic inventories.

PROCESSES IN FOREST CANOPIES

These are the most difficult to study without access methods that permit frequent returns to monitoring stations, because they require information about both sessile and mobile organisms, as well as interactions between the two groups. Reproductive biology is predominantly a canopy phenomenon in forests, although the pollinators and dynamics of flowering and fruiting in tropical trees is relatively unexplored (but see Mehra and Bawa 1968; Murawski 1995). Herbivory and insect–plant interactions have been quantified in several forests, using a combination of access methods, including SRTs, hot-air balloons, walkways, cherry pickers, and rafts (reviewed in Lowman 2001; Lowman 2009b). The distribution of insect herbivores is directly correlated to the location of their "salad bar" and the most desirable foliage to consume (Lowman 1982). Herbivores consume significantly less foliage in the upper crowns (sun leaves) as compared to the lower crowns (shade leaves), but young leaves (especially in the shade) are often completely consumed (Coley 1983; Lowman 1984; Lowman 1995). Foliage (a.k.a. leaf area) in tropical forests is most dense just below the upper canopy. Differences in herbivory levels can arise from artifacts of sampling, although canopy access has increased the accuracy of results, since the sun leaves can be measured in addition to the understory (reviewed in Lowman and Rinker 2004; Lowman and Wittman 1996; Schowalter and Lowman 1998, Table 3.2).

Access to tree crowns has stimulated studies of canopy-nutrient cycling, particularly with regard to epiphytes (Cardelús and Mack 2010). As tropical forests are cleared, the diversity of organisms supported by epiphytic communities becomes endangered or extinct (Cardelús et al. 2006). With predictions of warmer, drier climatic conditions, epiphytes are beginning to migrate up slopes along elevational gradients in the tropics, and they may eventually become extinct, according to models (Colwell et al. 2008). Other processes, such as photosynthesis, are more challenging to quantify for epiphytes or other canopy foliage, but construction cranes are providing important permanent sampling platforms for such measurements (e.g., Mulkey et al. 1996.). For the most part, the interactions of canopy processes remain unknown. Initial studies linking phenology of herbivory to decomposition on the rain forest floor of Puerto Rico relied mainly on SRT (Hunter et al. 2001). But links and synergistic outcomes have not been well quantified for any forest canopy processes, despite their emerging importance for

TABLE 3.2. Herbivory Levels (Percent Leaf Area Missing from Foliage-Feeders) in Forests of the World, as Measured by Limited Access to the Understory versus Canopy Access over Long-Term Field Sampling Methodologies (See also Lowman 1997)

Location	Herbivory (Percent Leaf Area Eaten / Year)	
	No Canopy Access	*With Canopy Access*
Belize (Central America)	7.0	—
Panama (Central America)	7.9	30
Peru (South America)	13.7	32
Cameroon	8.5	—
Australia		
Tropical	7.3	20
Subtropical	6.9	16
Montane	7.9	26
Dry Forest	7.3	15–300
Temperate deciduous	7.1	15.2

global health. For example, how does the seasonality of nutrient cycling affect pollination? How does herbivory affect tree growth and phenology and ultimately forest health? How is carbon sequestration affected by herbivory? Extrapolation of small-scale data collection to large-scale ecosystem applications is untested for most aspects of ecological processes.

A CHRONOLOGY OF IMPORTANT CANOPY
METHODS AND QUESTIONS THEY ADDRESS

Forest canopies long eluded scientists because of the logistical difficulties of reaching tree crowns and the subsequent challenges of sampling once there. Only in the last decade have field biologists designed an array of access methods that have inspired more extensive research on this unknown world of plants, insects, birds, mammals, and their interactions. Ideas about forest canopies had changed very little between Charles Darwin's time and the late 1970s, when biologists first adapted technical mountain-climbing hardware for ascending tall trees (Carroll 1979; Denison et al. 1972; Pike et al. 1977). SRTs employed versatile equipment to reach the midcanopy with ease and hang suspended on a rope to observe pollinators, epiphytes, herbivores, birds, monkeys, or other biological phenomena (reviewed in Lowman and Rinker 2004).

A number of critical issues have escalated global priorities in canopy research over the past two decades. First, as rain forests continue to decline due to human activities, the urgency of surveying the biodiversity of tree crowns before they disappear remains a major challenge for some researchers. Allegedly, many epiphytes and other plants, birds, mammals, and countless invertebrates inhabit the treetops (reviewed in May 2010), and an amazing number have escaped detection due to their aerial location. Many of these organisms are important not just as keystone species to the health of the rain forest ecosystem but also as sources of medicines, foods, or materials. Second, canopy processes are essential to life on our planet, and canopy organisms are part of the machinery that drives these important functions. As the resource economics of our planet become better understood and human health and ecosystem health prove inextricably linked, the rain forest canopy is emerging as a critical region that provides vital ecosystem services (reviewed in Laurance and Perez 2006; White et al. 2010). Forest canopies contribute directly to the global economy by providing oxygen (centers of photosynthesis), medicines, materials, foods, genetic libraries, nutrient cycling, carbon storage, climate stabilization including fresh water conservation, and a vast cultural heritage (reviewed in Hawken et al. 1999; White et al. 2010). And third, many researchers profess a simple curiosity to exploration of this previously inaccessible region of our planet. Relatively few unknown frontiers remain in the twenty-first century, but the treetops, ocean floor, and soil underfoot are still considered scientific "black boxes." The key to unlocking their mysteries depends on the design of new methods and technologies for access to measure these vast green landscapes (Fig. 3.2).

FIGURE 3.2.
Canopy methods used to study treetop biodiversity, physical attributes, and biological processes: A. Single rope techniques (India); B. towers (Costa Rica); C. construction cranes (Venezuela); and D. hot-air balloons (French Guiana).

Binoculars and telescopes were probably the first tools for canopy exploration. Alfred Wallace was one of the first naturalists to publish his observations of forest canopies (albeit from ground-based perceptions): "Overhead, at a height, perhaps, of a hundred feet, is an almost unbroken canopy of foliage formed by the meeting together of these great trees and their interlacing branches; and this canopy is usually so dense that but an indistinct glimmer of the sky is to be seen, and even the intense tropical sunlight only penetrates to the ground subdued and broken up into scattered fragments . . . it is a world in which man seems an intruder, and where he feels overwhelmed" (Wallace 1878).

Charles Darwin, less than a decade later but also in the nineteenth century, looked into the tropical rain forest foliage and exclaimed:

> Delight itself . . . is a weak term to express the feelings of a naturalist who, for the first time, has wandered by himself in a Brazilian forest. The elegance of the grasses, the novelty of the parasitical plants, the beauty of the flowers, the glossy green of the foliage, but above all the general luxuriance of the vegetation, filled me with admiration. A most paradoxical mixture of sound and silence pervades the shady parts of the wood. The noise from the insects is so loud, that it may be heard even in a vessel anchored several hundred yards from the shore; yet within the recesses of the forests a universal silence appears to reign. To a person fond of natural history, such a day as this brings with it a deeper pleasure than he can ever hope to experience again. (Darwin 1883)

During this period, plant collecting became a popular hobby and vocation—especially the sale of orchids. Early plant collectors not only contributed to our scientific knowledge of tropical forests, but they also led to the decline or even demise of some desirable species (reviewed in Hingston 1932; Moffett and Lowman 1995). Albert Millican's (1891) adventures in the Andes Mountains of Colombia were typical of this gold-rush of exploration for commercial trade: "After two months' work we had secured about ten thousand plants (of *Odontoglossum crispum*), cutting down to obtain these some four thousand trees, moving our camp as the plants became exhausted in the vicinity" (151).

William Allee (1926) published the first quantified measurements describing the canopy environment in Panama. The first recorded event where a human entered the canopy (rather than cut it down to ground level) was only three years later, when Major Hingston and colleagues erected an *observation platform* in British Guiana to hang baited traps for canopy organisms. Sadly, no data were published, but the chronology of canopy access begins with these observations of pioneer canopy scientists during the 1920s. During the 1940s, Charles Beebe (1949) used *rope ladders*, hoisted into the canopy on lines shot over branches, and brought the excitement of canopy biodiversity research to the general public with his popular book, *High Jungle*.

In the 1950s, a *steel tower* was constructed in Mpanga Forest Reserve in Uganda to study gradients from the forest floor to the canopy. Towers provided access to monitor insect vectors of human diseases, which remains the first (and landmark) applied biological study conducted in the forest canopy (Haddow et al. 1961). In the 1960s, *wooden and aluminum ladders* were also used for studies of canopy vertebrates, such as Robert Mc-Clure's (1966) studies of tree phenology and animal visitors. Ladders were used later for determining the pollinators of Dipterocarp trees (Gunatilleke and Gunatilleke 1996), for studying canopies in the Himalayas (Mehra and Bawa 1968; Mehra and Bawa 1969), and for surveying epiphytes in low-lying Florida oak hammocks (Doblecki and Lowman 1993). Other mid-twentieth-century tools, most of which were ground based, included training monkeys (Corner 1992); pruning poles (Crossley et al. 1976; Parker et al. 1993; Schowalter et al. 1991); slingshots, rifles, or insecticides to knock down samples (Erwin 1982; Lowman 1982); or hoisting sampling devices into the treetops, such as seed traps, butterfly baits, or small mammal traps (reviewed in Lowman and Moffett 1993; see also Chapter 5). Litter traps (see also Chapter 7) are one important example of methods that collect canopy material on the forest floor and are still used successfully for productivity estimates (see Lowman 1985; Odum and Ruiz-Reyes 1970; Schowalter et al. 1991).

The late 1970s represented the era of direct access into forest canopies using SRTs. New hardware designed for caving and mountaineering was adapted for tree climbing and made these techniques relatively safe and inexpensive. In addition, the portability of SRTs for direct canopy access allowed graduate students and others with a modest budget to survey life at the top and replicate their data collection between and among different tree canopies. Don Perry (1978) first used SRTs at LaSelva in Costa Rica, examining the ecology of a Ceiba tree. At the same time (but without the luxury of e-mail communication between the two researchers), Meg Lowman developed SRTs in Australian rain forest canopies, also starting in 1978, by adapting caving hardware from the Sydney University Caving Club (reviewed in Lowman 1999). Perry went on to develop the canopy web, the aerial tram, and other methods that were creative extensions of a rope system (Perry 1986; Perry 1995). Lowman went on to design canopy walkways that allowed teams of volunteers or scientists to work simultaneously in the treetops, because she relied on groups of citizen scientists such as Earthwatch teams to survey arthropod biodiversity in forest canopies (Lowman 1985; Lowman 1988; Lowman 2009b; Lowman and Bouricius 1995). SRTs were not effective for large groups of researchers, or to reach the outer perimeter of tree crowns, since the ropes had to be looped over sturdy branches, usually close to the tree trunk. SRTs are not as safe as access techniques that have a secondary safety mechanism. Subsequently, double-rope techniques (DRTs) were developed because this offered a duplication of efforts important to the safety of a user (http://www.newtribe.org).

Specific methods based on adaptations of rope systems were also developed in the 1980s for specific hypotheses. The peconha (climbing loop of webbing held around a tree trunk) was useful for studies in relatively small-crowned trees with straight trunks, including surveys of small mammals in Brazil's forest fragmentation plots (Malcolm 1991). Climbing spikes were commonly used to survey Brazilian nut trees (see Mori 1984), although they are no longer advocated due to damage inflicted on tree trunks that can lead to inoculation sites for tree pathogens and epiphyte mortality. Tree bicycles were of limited use to sample flowers or fruits for taxonomic surveys (Gentry 1991). To access the outer foliage of canopy trees, Appanah and Ashton designed the canopy boom, a horizontal bar with a bosun's chair at one end, which could be swung around into the canopy away from the tree trunk. In Pasoh, Malaysia, booms supplemented the ladder methodologies (described previously) to help solve the mystery of dipterocarp pollination (Appanah and Chan 1981).

In the early 1980s, Terry Erwin (1982) revolutionized our estimates of biodiversity by introducing the insecticidal fogging apparatus into canopy research. This method frequently required single rope access into tree canopies, to create the hoist for fogging apparatus. Although this sampling technical results in mortality of most arthropods within a vertical column of forest, it remains the only means to create a comprehensive collection of insect biodiversity at one point in time. By misting the treetop of a *Luehea seemanii* in Panama, Erwin collected the rain of insects and counted the diversity of species, especially beetles. His extrapolations raised our estimates of global biodiversity from almost 10 million to more than 30 million. Fogging continues to be utilized in a limited fashion by rain forest biologists who need to estimate the diversity of life in the treetops. Collectively, these sampling techniques in the 1980s moved canopy science from a "sense of wonder" and exploration to a more rigorous science whereby hypotheses were tested and vast databases were filled. Others adapted fogging (or misting) to their own sites and conditions, leading to some important comparative studies of arthropod diversity in tropical forests throughout the world (e.g., Meg Lowman and Roger Kitching in Australia; Nigel Stork in Malaysia; Richard Southwood in United Kingdom).

In the mid-1980s, biologists began utilizing combinations of several canopy access techniques, still primarily to survey biodiversity. Walkways were often used in combination with SRTs and perhaps canopy booms, ladders, cherry pickers, or other creative means (reviewed in Lowman and Moffett 1993). In 1985, engineer Ilar Muul built the first tree-supported canopy walkway at Lambir National Park, Malaysia, anchored on tree crowns. At the same time, Lowman and eco-tourist lodge owner Peter O'Reilly designed the first pole-supported structure (not live-tree cabling) near Lamington National Park in Queensland, Australia, intended both for ecotourism as well as to facilitate Lowman's Earthwatch teams, who were surveying insect herbivores in tree canopies (reviewed in Lowman 1999). Walkways represented an advance over SRTs because groups could work together, especially during relatively inclement weather conditions, to collect data. By contrast, SRTs allow only one person to utilize a rope on a single vertical transect, and work cannot be conducted during inclement weather due to safety concerns. Use of canopy walkways burgeoned in the early 1990s, with the modular construction design of Lowman and Bart Bouricius (1995) that could utilize pole support, tree support, or a combination of both (http://www.canopyconstruction.com). Since then, canopy walkways and ladders used in conjunction with climbing ropes and other tools have become as popular as permanent canopy field sites (reviewed in Lowman and Wittman 1996) and for ecotourism that inspires forest conservation by providing a revenue stream to indigenous forest dwellers (Lowman 2009a).

The last chapter in the development of the toolkit for canopy research involves the addition of integrated, collaborative research. The methods of the 1970s and 1980s (e.g., SRT, booms, cherry pickers, scaffolding, ladders, and the earlier canopy platforms [premodular construction]) were relatively limited in scope, favoring solo work or small groups. Two scientists expanded the scope of canopy access tools with their creative genius on separate continents: Francis Halle of the Institut de Botanique in Montpelier, France, and Alan Smith of the Smithsonian Tropical Research Institution, Panama. Halle designed a colorful hot-air balloon, *Radeau des Cimes*, and first deployed it with a team of more than twenty biologists in French Guiana in 1987 (Hallé 2002). Its inflatable raft was 27 meters in diameter and formed a platform on top of the forest canopy that was utilized as a base camp for research on the uppermost canopy. The dirigible, or hot air balloon, was used in conjunction with the raft, serving to move the raft to new positions throughout the jungle and also to give researchers access into the above-canopy atmosphere for studies of the canopy–air interface. In 1991 in Cameroon, Africa, the *Radeau des Cimes* expedition team pioneered a new canopy technique called the sled, or skimmer (Lowman et al. 1993). This small (5 meter) equilateral, triangular miniraft was towed across the canopy by the dirigible, similar to a boat with a trawling apparatus in the water column of the ocean. It facilitated the rapid collection of canopy leaves, flowers, vines, and epiphytes, as well as their pollinators and herbivores. In Madagascar in 2001, Halle's team launched a new device that was essentially an individual canopy cell within the crown of one tree, whereby researchers could be dropped off by the balloon for temporary residence inside the metal frame of the canopy camp.

Similar to the collaborative aspects of the balloon expeditions, the first canopy construction crane was erected by Smith, tropical botanist at the Smithsonian Tropical Research Institute in Panama (reviewed in Parker et al. 1992). In 1990, a 40-meter-tall crane was erected in a tropical dry forest outside Panama City, Panama. Cranes are expensive to install and operate (usually ranging from $1 million to $5 million), but they offer unparalleled access to the uppermost canopy as well as to any section of the understory that is within reach of the crane jib and have the additional advantage of providing access to canopies beyond the footprint of the tower. Since the success of the first crane in Panama, approximately ten cranes are situated around the globe, including North America, Japan, Australia, Switzerland, Germany, Malaysia, and (formerly) Venezuela (reviewed in Mitchell et al. 2002). With one permanent site, diverse researchers can work collaboratively and comparatively, both within and between crane sites (reviewed in Stork 2007).

Continuous remote-sensing capability provided by the variety of satellite-based sensors orbiting the Earth has also revolutionized canopy research (see also Chapters 2 and 3). Images taken by infrared, ultraviolet, and other sensors can be obtained from various government agencies and provide data not only on canopy cover but also on species composition, tree condition, and foliar and aerosol concentrations of various chemicals (e.g., Porder et al. 2005). In the next decade, new devices such as tall towers (Amazon Tall Tower Observatory [ATTA]) and Google's launch of new imaging software to assess forest canopy cover represent emerging technologies that will facilitate more extensive and accurate research on global forest canopies.

FUTURE DIRECTIONS

Given this arsenal of canopy tools developed during the late twentieth century (Fig. 3.2), what is next? The biggest priority in canopy research remains the challenge of continuous, long-term access for teams of scientists. At the first international canopy conference held in Sarasota Florida in 1994, Andrew Mitchell, now director of the Global Canopy Programme (http://www.globalcanopyprogramme.com), aspired to create the most ambitious canopy tool ever: biotopia. This concept would integrate several field methods together, including cranes, walkways, canopy rafts, towers, and ropes, essentially comprising a field station dedicated to canopy research. This notion was subsequently endorsed by E. O. Wilson and other prominent biologists who believe that a large, long-term, well-funded effort to document terrestrial biodiversity is long overdue. Like the particle accelerator or large-scale programs for NASA to study regions of outer space, a canopy field station or biotopia would allow unprecedented research throughout Earth's forests, still a relatively understudied component of our planet.

Now that scientists have creatively designed a toolkit for canopy access, the real challenges lie ahead. Canopy organisms—both mobile and sessile—must be surveyed and their roles measured. Canopy processes must be sampled with respect to the complex differences in light, height, tree species, and seasonality. Establishing rigorous sampling techniques and conducting long-term research while dangling from a rope remain ambitious objectives. The next ten years are critical, as scientists attempt to classify the biodiversity and ecology of forest canopies before exceeding the tipping points of habitat fragmentation and climate change. Exciting new directions exist in the near future—the extrapolation from leaf to canopy, from organisms to populations, from flower to entire crown, from seedling mortality to recruitment patterns throughout forest patches, from light levels in small gaps to photosynthesis of entire stands, and most importantly, from canopy to whole forest. Canopy access techniques will provide invaluable access to this exciting region of our planet and hopefully yield results that will facilitate the implementation of sound conservation practices. The emerging links between forest canopies and human health, as well as resource economics at a global scale, will increasingly drive new directions for canopy research.

SUGGESTED READING

Laurance, W. F., and C. A. Perez, eds. 2006. *Emerging threats to tropical forests.* Chicago: University of Chicago Press.

Leather, S. R., ed. 2005. *Insect sampling in forest ecosystems.* Malden, MA: Blackwell Science.

Linsenmair, K. E., A. J. Davis, B. Fiala, M. R. Speight, eds. 2001. *Tropical forest canopies: Ecology and management.* Dordrecht, Netherlands: Kluwer Academic.

Lowman, M. D. 1999. *Life in the treetops.* New Haven, CT: Yale University Press.

Lowman, M. D., E. Burgess, and J. Burgess. 2006. *It's a jungle up there.* New Haven, CT: Yale University Press.

Lowman, M. D., and H. B. Rinker, eds. 2004. *Forest canopies.* 2nd edition. San Diego, CA: Elsevier/Academic Press.

Mitchell, A. W. 1986. *The enchanted canopy.* Glasgow, Scotland: Williams Collins and Sons.

Mitchell, A. W., K. Secoy, and T. Jackson, eds. 2002. *The global canopy handbook.* Oxford, UK: Global Canopy Programme.

Moffett, M. W. 1993. *The high frontier.* Cambridge, MA: Harvard University Press.

Nadkarni, N. 2008. *Between earth and sky.* San Diego CA: University of California Press.

Perry, D. 1986. *Life above the jungle floor.* New York: Simon and Schuster.

4

FOREST STRUCTURE
AND SAMPLING UNITS

Canopy structure reflects the three-dimensional framework of boles and branches and the distribution of foliage that represent the engine for capture and processing of energy and atmospheric nutrients, processes fundamental to sustainability of forest ecosystems and the services they provide. Furthermore, canopy structure establishes the framework of support and resources for associated canopy flora and fauna, as well as the template of sample units for canopy research. Canopy structure determines rates and pathways of fluxes of carbon, nutrients, and organic material from the canopy to the atmosphere, forest floor, and associated stream systems. Consequently, adequate representation of forest structure is a primary goal of canopy research.

Forest canopies have challenged description of three-dimensional structure, especially in tall, densely packed canopies such as tropical rain forests. Difficulty in accessing forest canopies has been a major impediment that has generated much of the creativity in canopy access methods described in Chapter 3. The extreme three-dimensional structure of trees and forests also has generated problems in selecting appropriate sampling units for particular objectives and for placement and monitoring of instruments, especially as compared to relatively two-dimensional ecosystems such as rocky intertidal shorelines (Lowman and Moffett 1993; Murray et al. 2006). Taller or denser canopies require different experimental approaches than do shorter or sparser canopies. The results of many canopy studies have been compromised by lack of adequate independent replication of sampling units. For example, a comparison of two tree crowns accessible from a tower does not provide any error degrees of freedom for statistical comparison. Manipulation of canopy variables, especially in older forests, is particularly difficult, limiting advances in understanding canopy biology. Finally, the variety of confounding variables associated with complex three-dimensional structure and with repeated sampling of accessible experimental units requires caution in the choice of statistical analyses.

Forest types and ages vary in their vertical and horizontal structure, as described in Chapter 2, reflecting differences in forest development under different environmental conditions. This chapter describes measurement of vertical and horizontal variation in canopy structure at the stand level and addresses criteria and considerations for selection of sample units and experimental design in forest canopies. Experimental design will determine selection of the most appropriate access and sampling methods, as described in Chapters 3 and 5.

CANOPY STRUCTURAL VARIABLES

Canopy structure reflects interaction between forest development and environmental conditions and determines foliage distribution, photosynthetic efficiency, and canopy–atmosphere interaction (e.g., Baldocchi 2008). Forest canopies are not uniform surfaces but rather show considerable variation in height among trees, as well as spacing among individual crowns at any given height. As a result, the canopy surface has a three-dimensional topography that varies among forest types and ages. Variation among crowns in branch configuration and foliage form and density, among other parameters, also affect canopy structure. Canopy cover and within-canopy variation in biodiversity, environmental conditions, and ecosystem processes are determined by canopy structure. The following components of canopy structure affect canopy environment and function and should be measured, as appropriate.

VERTICAL AND HORIZONTAL VARIATION

Vertical and horizontal structure reflects canopy height and the extent of subcanopy and gap formation. Vertical and horizontal structure vary widely, from relatively flat surfaces and uniform height of younger forests dominated by pioneer species to complex topography and multicanopied structure of older forests that reflect the influence of disturbances and stage of succession (Sillett and Van Pelt 2007; Song et al. 2004; Van Pelt and Nadkarni 2004). The canopy surface of old forests typically shows peaks formed by emergent crowns of dominant trees; valleys and troughs formed by shorter canopies and gaps, resulting from tree falls; and flatter surfaces formed by patches of codominant trees of uniform age and height. The surface topography of forest canopies influences a variety of processes, including the turbulence of air movement, efficiency of interception of light and precipitation, and gas exchange (Finnigan et al. 2009; Su et al. 2008; Turner et al. 2005; Turner et al. 2007; see Chapter 6).

Canopy height and density determine the degree to which distinct vertical gradients in light penetration, temperature, and relative humidity develop (Madigosky 2004; Parker 1995). Taller and/or denser canopies have greater capacity to modify environmental conditions and show more distinct gradients than do shorter and/or more open canopies. Older forests tend toward greater complexity in canopy height and crown form as a result of bole and branch breakage, tree fall, and subsequent replacement of gaps by younger trees, leading to greater variation in vertical and horizontal gradients in environmental conditions within the canopy (Fig. 4.1).

Tropical forest canopy surfaces can be represented as gently rounded mounds, due to the umbrella-shaped crowns of the dominant angiosperms (Fig. 4.1A), whereas boreal forest canopy surfaces are represented as sharper peaks, due to the more columnar and/or pointed crowns of gymnosperms (Fig. 4.1B; see also Chapter 2). Sunlight from directly above tropical forests penetrates deeply through shallow crowns, permitting development of multistoried canopies composed of a diversity of overstory and understory species, including trees no taller than 1–2 meters at maturity that persist in the low-light environment near the forest floor. By contrast, sunlight from lower in the sky in boreal forests penetrates primarily from the side, favoring deep, narrow crowns on overstory trees and limited understory development (Terborgh 1985).

Vertical changes in canopy structure develop as a result of multicanopy formation by overstory and understory tree species in tropical and late-successional forests and/or as a result of temperature and moisture gradients in deep canopies. Increasing solar exposure and temperature and decreasing relative humidity and CO_2 concentration with increasing canopy height lead to changes in leaf morphology and orientation relative to the sun angle (Gutschick 1984; Gutschick 1999; Gutschick and Wiegel 1988). Leaves at the canopy surface typically are smaller, thicker, and highly sclerotized to protect the leaf from ultraviolet damage (typical of leaves in more arid environments), whereas leaves deeper in the canopy are larger, thinner, and less sclerotized (Gutschick 1999; Lowman 1985). Such vertical layering results in differences among canopy layers in contributions to ecosystem productivity and carbon exchange with the atmosphere (Misson et al. 2007). Furthermore, the vertical variation in canopy environmental conditions results in trends with height in epiphyte and animal abundances and diversity (Basset et al. 2007; Fortin and Mauffette 2002; Schowalter and Ganio 1998; see Chapter 5). Vertical layering may not produce discrete strata, as suggested by much canopy research (Parker and Brown 2000), but vertical variation in canopy variables must be represented in canopy studies.

Although canopy studies typically focus on living trees, dead trees (or portions of trees) become an increasingly important component of the canopy system as forests

FIGURE 4.1.
Crown form of (A) tropical trees and (B) boreal trees. Note the rounded, umbrella form of the tropical crowns; the narrow, conical form of the boreal crowns; and the penetration of sunlight from the side in the boreal forest.

on the vertical surface, whereas more isolated crowns also are exposed on lateral surfaces in the direction of the sun, which can change diurnally and seasonally, or wind. Increasing crown isolation permits greater penetration of light to subcanopy layers and the forest floor. Storm damage to crowns in contiguous canopy might be limited to branch breakage along the upper surface or where adjacent crowns abrade each other, whereas more isolated or exposed crowns suffer greater breakage, lean into neighboring trees, or even fall to the ground, depending on wind speed and degree of crown resistance to airflow (Moore and Maguire 2005; Stork 2007; Zimmerman et al. 1994).

Tree position relative to other trees also determines canopy connectivity for associated organisms. For example, crowns surrounded by trees of the same species are more likely to be contacted by propagules of epiphytes or parasites on neighboring trees or by foraging or dispersing herbivores and other organisms (Shaw et al. 2005). Crowns in contact with other crowns are more likely to serve as corridors for movement through the canopy than more isolated crowns. Frequently, crowns of neighboring trees at the same canopy level tend to show some degree of separation, a condition known as "crown shyness" (Fig. 4.2), as a result of breakage of branch ends as abutting crowns abrade each other (Fish et al. 2006; Putz et al. 1984; Rebertus 1988). The space between crowns permits penetration of light, precipitation, and wind; creates eddy turbulence; and provides flyways for birds and other flying animals. However, many nonflying animals are able to leap across the space between crowns (Dial et al. 2004).

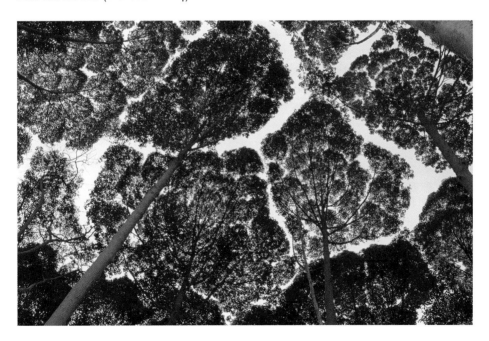

FIGURE 4.2.
Crown shyness in a tropical rain forest.
Photo by Mark Moffett.

Finally, crown spacing determines canopy coverage of the understory and forest floor. The extent of canopy coverage affects the degree to which light, water, and wind penetrate to the ground between crowns. These factors determine albedo (the reflection coefficient, or reflecting power, of a surface), soil temperature and moisture, raindrop impact and erosion, and evaporation, all of which affect the canopy environment, as well as local and regional climatic conditions (Foley et al. 2003; see Chapter 7).

FOLIAGE DISTRIBUTION

Distribution of foliage is a primary aspect of forest structure that affects rates of photosynthesis, evapotranspiration, net gas exchange, and interception of precipitation, particulates, and aerosols (Asner and Martin 2011; Dominy et al. 2003; Ellsworth and Reich 1993; Fyllas et al. 2009; Gutschick 1984; Harley et al. 1996). The pool of foliage is composed of different tree species, which differ in size, shape, and various chemical components, and different configurations and chemical composition within trees, based on availability of light, water, and nutrients, as well as foliage of associated epiphytes.

Obvious differences in foliage distribution arise from differences in leaf size, shape, thickness, and within-branch density among tree species (e.g., between conifers, with dense needle-shaped foliage, and broad-leaved angiosperms), among broad-leaved species with simple versus compound leaves, and between deciduous trees, with foliage only during the growing season, and evergreens, which retain foliage year round and often for several years. Foliage size, shape, thickness, and density represent important tradeoffs among photosynthetic efficiency, energy and nutrient investment, and ease of replacement (Gutschick 1999; Gutschick and Wiegel 1988). Deciduous trees typically retrieve nutrients, especially nitrogen, prior to senescence (Marschner 1995), resulting in different qualities of litterfall contributing to canopy–forest floor interaction (see Chapter 7).

The arrangement of leaves along branches also varies among tree species and reflects specific adaptations to optimize photosynthetic efficiency. Trees with long-lived leaves that can maintain photosynthesis under diffuse light may retain leaves along branch lengths (e.g., Douglas-fir), whereas trees with short-lived leaves or leaves that require full sun may retain only a cluster of leaves at the exposed ends of branches (e.g., *Cecropia* sp.). Leaf distribution and photosynthetic efficiency also reflect leaf angle relative to branch angle and direction of sunlight.

The arrangement of tree crowns of various species in three-dimensional canopy spaces also affects foliage distribution. Some crowns have denser foliage (higher leaf-area index) than others.

REPRODUCTIVE STRUCTURES

Reproductive structures form an important component of the canopy system, including both host trees and layers of epiphytes, epiphylly, and vines. The type and distribution of reproductive structures depends on tree species, position in the canopy, and major pollination and seed dispersal mechanisms. Flowers, fruits, and seeds represent important food resources for a variety of canopy vertebrates and invertebrates, and the spatial and seasonal distribution of these resources strongly influence animal foraging activity (e.g., Roubik 1993; Schowalter 1993).

Major differences in floral type and distribution are obvious between temperate and tropical forests. Temperate forests are dominated by wind-pollinated species, whose reproductive structures are catkins on the ends of branches, usually high in the canopy. This pollination system is favored in forests with relatively low diversity and high probability that flowers will be fertilized by conspecific pollen dispersed randomly on the airstream (Moldenke 1976).

By contrast, tropical forests have high diversity and low probability that flowers will be fertilized by conspecific pollen dispersed in random fashion. Pollination of tropical trees is primarily by invertebrate, avian, and bat pollinators, and floral color, structure, and nectar resources are adapted to attract specialist pollinators, whose fidelity to particular plant species ensures pollination (Roubik 1993). Overstory trees generally are characterized by well-illuminated, brightly colored flowers, held above the canopy to attract pollinators over a wide area, and brief, highly synchronized flowering periods within plant species (Appanah 1990). Dominant pollinators are bees, such as *Apis dorsata*, and trapliners such as carpenter bees, birds, and bats. Some overstory dipterocarps in the genera *Shorea*, *Hopea*, and *Dipterocarpus* have tiny flowers with limited nectar rewards, flower nocturnally, and are pollinated by thrips and other tiny, flower-feeding insects (Appanah 1990). Some plant species representing various canopy strata are cauliflorous (i.e., they produce flowers along the trunk or main branches). These flowers typically are large, or small and clumped, pale colored, odiferous; show a brief, highly synchronized flowering period; and are pollinated by understory and overstory insects, birds, and bats (Appanah 1990).

Understory trees in both forest types occupy canopy positions with low visibility and airflow and depend on animal, primarily insect, pollinators to transport pollen among conspecific trees. Flowers generally are white or pale colored, offer limited nectar rewards, often produce floral scents that resemble dung or carrion to attract flies and beetles characteristic of the forest floor, and are open over extended periods to ensure pollination by infrequent visitors. They are pollinated by nonselective beetles and flies or by nonspecific trapliners—that is, species that revisit particular plants along an established circuit.

Distribution of fruits and seeds in the canopy reflects the distribution of floral structures. Fruit and seed size reflect the primary dispersal mechanism. Temperate overstory trees typically produce lightweight, often winged seeds that can be carried long distances by wind. Some understory trees produce fruits that attract animal dispersers. Again, tropical overstory trees more often produce fruits to attract animal dispersers that transport seeds from the low-light environment in the vicinity of the parent to more favorable locations for germination. Larger fruits attract large mammals, whereas smaller fruits attract small mammals, reptiles, and birds, which disperse seeds more widely.

A number of other canopy components provide habitat or food resources for associated organisms. Among these are epiphytic mats, suspended soils, and aquatic habitats.

As trees age, the increasing size of branches and bark crevices leads to accumulation of organic and inorganic debris within the canopy. Soil formation provides substrate for the development of plant and soil/litter communities normally associated with the forest floor (e.g., Coxson and Nadkarni 1995; Fonte and Schowalter 2004b; Winchester 2006). Plants colonizing these perched soils can be understory species gaining a foothold in the canopy, hemi-epiphytes that colonize overstory trees and eventually kill and replace them, or other canopy tree species (Sillett and van Pelt 2007). However, given differences in temperature and relative humidity between the forest floor and upper canopy, soil communities typically are composed of arboreal specialists. For example, Lindo and Winchester (2008; 2009) reported significant differences in the species composition of oribatid mite assemblages between the forest floor and canopy of old-growth conifer forests on Vancouver Island, British Columbia.

Tree holes form as a result of branch breakage, bole decay, and/or excavation by animals (e.g., insects, woodpeckers, and squirrels). Some tree holes fill with water and provide suspended aquatic habitats of variable persistence for specialized aquatic organisms (Yanoviak 1999; 2001). Water collecting in the bowls formed by birdnest ferns and bromeliads also provide important suspended temporary pools for aquatic organisms. The colonization of tree holes by predators and the amount of organic litter accumulation affect community development (Yanoviak 2001).

Epiphytic mats develop as various lichens, mosses, and so on colonize suitable portions of boles and branches (Yanoviak et al. 2007). These mats contribute to photosynthesis, carbon exchange, water retention, and soil accumulation in the canopy.

Animals may form or modify other structures in canopies as a result of nest building or other activities. Bees, ants, and termites construct arboreal nests from sediments and/or organic matter, which increases the complexity of canopy structure (see Chapter 5).

MEASUREMENT METHODS

The earliest method for measuring tree and forest structural parameters applied geometric principles—that is, the calculation of tree height from the distance of an observer from the tree (base of a right triangle) and the angle of a sight line to the top of the tree, measured using a protractor. Crown shape was measured using similar calculations (Horn 1971). However, this method requires that the base and top of the tree be visible to the observer—a condition rarely met in dense forests. Furthermore, this method was clearly labor intensive for measuring heights of multiple trees and thereby limited the amount of data that could be collected from forests. Foliage density can be measured most easily by photometric methods. Foliage height profiles were measured by extending a long pole and counting the number of leaves that touch the pole within specified height increments (e.g., Schemske and Brokaw 1981; Wunderle and Waide 1993). This technique has limited application in forests with tall or dense foliage height profiles.

Among the earliest efforts to define and measure crown architecture were those of Addor et al. (1970). They defined a number of variables, including crown branching types, diameter and depth, height, angle and length of branches, and size and shape of leaves. Measurement of three-dimensional canopy architecture has become easier with modern access and GPS methods. Towers and cranes permit the convenient measurement of overall tree geometry and access to within-canopy structures. From positions within crowns, the detailed measurement of three-dimensional structure becomes possible.

Crown mapping provides the context for measurements of canopy microclimatic conditions, resource distribution, community structure, and canopy processes (Dial et al. 2004; Sillett and Van Pelt 2007). Ideally, tools for measuring these variables would be positioned randomly within crowns, but this rarely is possible. The location of measurements often depends on branch structure, in terms of both orientation and support, particularly for heavy equipment. Towers, cranes, or other support structures that may facilitate canopy access and transport of heavy equipment also preclude random location of measurements, although the x-, y-, and z-coordinates for gondola position provided by canopy cranes can be used for sampling at random coordinates accessible to the hook. Therefore, devices anchored to towers or cranes will be most useful for measuring variables on the periphery of tree crowns, but access to the interior of crowns is limited. Devices that require support by the bole or branches limit measurement at the crown periphery. Sillett and Van Pelt (2007) described placement of various monitoring devices within crowns of redwood trees.

Various methods have been developed for crown mapping, all aided by recent advances in three-dimensional geopositioning tools. However, while crown mapping may be conducted most conveniently from towers or cranes, adequate access to the interior of the crown for description of branch network structure still requires climbing techniques. Ishii and Wilson (2001) climbed 50- to 62-meter-tall, 500-year-old Douglas-fir trees at the Wind River Canopy Crane Research Facility in Washington state, using the single-rope technique. All primary branches were numbered and recorded as live or dead. Branch height above ground, diameter at the bole, and length to the farthest foliated section were measured to describe tree crown structure. Similarly, Sillett and Van Pelt climbed 100-meter redwood trees in California and mapped the complex reiterated crown structure. Dial et al. (2004) established horizontal transects between emergent trees at sites in Alaska and Washington state, Costa Rica, Borneo, and Australia. Vertical transect ropes were dropped every 5–20 meters along the horizontal transect, and distances to the nearest objects at 1–2 meter height increments were recorded for equally spaced azimuths to characterize canopy form (Dial et al. 2004).

Distribution of leaf area and biomass within the crown can be measured from above or below the canopy, primarily using photographic methods. *Leaf area index* (LAI) is a fundamental measure of foliage distribution and density that has been used not only to describe canopy structure but also to represent a unit for describing densities of various canopy organisms. Leaf area can be measured by indirect methods such as canopy photography and plant canopy analyzation (which require data on relationships between light penetration and LAI) and by direct methods such as allometry (which requires the measurement of species-specific relationships between leaf area and bole diameter) and leaf collection in litterfall traps (which works best in deciduous forests). The earliest, and still widely used, method for measuring foliage density is measurement of the percentage of sky obstructed by foliage using photographic techniques (Horn 1971). Fish-eye photographs taken vertically from the ground or at designated heights can be analyzed digitally to measure LAI and foliage distribution and provide estimates of foliage biomass (Frazer et al. 2001; Uriarte et al. 2005). Remotely sensed images from above the canopy can be analyzed spectrometrically to measure canopy height, crown cover, foliage density profiles, LAI, foliage distribution, and so on (Gutschick 1996; Helmer et al. 2010; Turner et al. 2005). Light detection and ranging (LiDAR) equipment and moderate resolution imaging spectroradiometers (MODIS) are commonly used remote-sensing technologies for characterizing canopy structure (Chambers et al. 2007; Heinsch et al. 2006; Lefsky 2010; Means et al. 1999; Treuhaft et al. 2002; Treuhaft et al. 2003; Treuhaft et al. 2004; Turner et al. 2005). Ryu et al. (2009) compared two direct methods (litterfall, allometry) and five indirect methods (LI-COR LAI–2000 plant canopy analyzer, third-wave engineering tracing radiation and architecture of canopies [TRAC] instrument, digital hemispheric photography, digital cover photography, and traversing radiometer system). They found that the combination of digital canopy photography and LAI–2000 could provide spatially representative LAI, gap fraction (a measure of light penetration at a particular angle as a proportion of leaf orientation at that angle), and element dumping index (a measure of leaf distribution).

Crown spacing and canopy cover can be mapped most easily using aerial photography or remote sensing, GPS, and geographic information system (GIS) tools (Anzures-Dadda and Manson 2007; Song et al. 2004; Treuhaft et al. 2003; Treuhaft et al. 2004; Van Pelt and Nadkarni 2004; Wang et al. 2006). Bole coordinates and distances to nearest neighbors can be mapped from the ground to provide general location information. Three-dimensional crown structure can be mapped within the canopy by recording x-, y-, and z-coordinates of points on crown perimeters that are accessible from various vertical transects (Dial et al. 2004) or by recording branch height and distance and azimuth to the main trunk (Sillett and Van Pelt 2007). Crown damage following disturbance can be measured from within or above the crown, but safety issues may require that crown damage be assessed from the ground using photographic techniques.

Abundance and distribution of fruits and seeds can be measured in the canopy as part of the three-dimensional description methods described previously or by using seed traps distributed on the forest floor. Traps distributed along radii from the bole to the outer periphery of the crown provide data on the horizontal distribution of seed production but not its three-dimensional distribution in the crown. Other canopy structural components (e.g., tree holes and epiphyte mats) can also be measured as part of other measurements of three-dimensional canopy structure (as discussed previously).

SELECTION OF SAMPLE UNITS

Forest canopies can be studied at multiple scales, depending on hypotheses and objectives. Scale of study can be individual leaves, individual trees, tree species, forest patches, or entire forests. Canopies are composed of the crowns (i.e., foliage and branches) of trees that compose the forest. Individual tree crowns are a network of radiating branches and foliage. Multiple crowns represent the diversity of vertical and horizontal structure of various tree species as influenced by resource (light, carbon, and nutrient) availability and competition with surrounding trees. Sample unit selection also depends on canopy structure. Young forests with higher stem density and shorter canopy may be conveniently studied in small plots (e.g., 50 x 50 meters containing multiple trees), whereas old forests with low stem density and taller canopy would require considerably larger plots (> 1 hectare) to include multiple trees.

For some objectives, measurements are most appropriate or accurate at the scale of a leaf, but results must be integrated at the canopy level. The following examples illustrate criteria for selection of sample unit.

Several methods have been developed to measure foliage density and light levels through the canopy (in order to estimate primary productivity; see Chapter 6). Prior to remote-sensing techniques, such measurements were confined to ground-based methods, including measurements from towers, making individual trees the sample unit (Horn 1971). Clearly, representation of foliage density throughout a crown is necessary to represent variation due to leaf position in the crown, position in the canopy, and so on, as these determine access to light, water, and nutrients. An average foliage density for each tree then could be multiplied by the density of trees to obtain foliage density per hectare, which could be multiplied by average rate of photosynthesis to estimate productivity. Such scaling up to describe whole-plant canopies requires computationally efficient models that incorporate relationships among values measured at each scale, accounting for variation in foliage density with light level at increasing depth through the canopy and variation in foliage density among overstory and understory tree species (Gutschick 1996; Horn 1971; Turner et al. 2007; see Chapter 6). Now the amount of vegetation can be quantified by remotely sensed spectral indices, such as the normalized difference (of infrared [IR] and red [R] radiances) vegetation index (NDVI), as

$$NDVI = (IR - R) / (IR + R),$$

but without local calibration of NDVI to foliage biomass or LAI, the accuracy of remotely sensed data cannot be determined (Gutschick 1996). Note that errors associated with variation in measurement at the scale of the leaf increase as the data are scaled up to estimates of number of leaves per tree or per hectare of forest.

For some tree species that sprout from roots (e.g., *Populus tremuloides* or *Robinia pseudoacacia*), multiple trees could represent modules of a particular genotype. Therefore, samples from multiple trees of such species would not necessarily represent independent replicates, if comparison of foliage density (for example) among tree species or forest types was an objective. Selection of appropriate sample units could be modified accordingly.

Individual branches can be used as sample units, for example, for measurement of differences in branch angle, foliage density, epiphyte density, or gleaning/foraging by animals among crown strata or tree species (e.g., Mooney 2007). However, because only the outer portion of branches is foliated for most tree species, many variables will have distinct values for proximal and distal portions. Furthermore, as described previously, branches from the same tree would not represent independent replicates for comparison of variables among tree species or forest types.

For most studies that address structure or function at the forest scale (e.g., variation in foliage density among tree species or forest types), sampling would represent multiple trees. However, although multiple measurements from individual trees provide more representative data, multiple measurements from an individual tree are not statistically independent for purposes of comparison or for calculating a value at the forest scale. Individual trees and tree species vary in structure as a result of differences in genome, exposure to light, water or nutrient availability, and other environmental factors, such as disease or disturbance. Therefore, measurements for each tree must be pooled, and means per tree become the unit for independent replication.

Remote sensing has provided a convenient means of measuring many canopy variables at stand, landscape, or regional scales. These integrated measures of canopy structure, albedo, gas exchange carbon flux, evapotranspiration, and so on should reduce errors associated with scaling up from leaf measurement and represent processes over the large areas necessary for evaluation of global changes (Gutschick 1996; Law 2005; Law et al. 2006; Lefsky 2010; Porder et al. 2005; Pypker et al. 2005; Treuhaft et al. 2003; Treuhaft et al. 2004). However, calibration of such measurements, to ensure accuracy, requires comparison with detailed measurements at leaf or tree scales (Gutschick 1996; Mariotto and Gutschick 2010).

EXPERIMENTAL MANIPULATION OF CANOPY STRUCTURE

Many studies of canopy biology do not require experimental manipulation. However, for definitive testing of hypotheses about factors affecting canopy structure or function or canopy effects on global processes, experimental manipulation is essential. Manipulation of forest canopies is obviously difficult because of limitations to access and maintenance of experimental conditions, and relatively few studies have employed manipulation for hypothesis testing.

Most canopy studies have exploited unplanned experimental designs resulting from natural disturbances or land-use practices (e.g., effects of forest harvest or fragmentation on canopy communities or processes; Anzures-Dadda and Manson 2007; Malcolm 1995; Schowalter 1995; Schowalter and Ganio 2003; Stouffer et al. 2009; Wunderle et al. 2006). Sufficient independent replication of disturbed and undisturbed sites has provided much of our knowledge of the effects of various disturbances on forest structure and its recovery. However, such "natural" experiments are subject to misinterpretation due to differential effects of disturbance or land-use change among forest structures that reflect underlying confounding variables, such as aspect, topographic position, or substrate conditions (Walker 1991).

In a number of studies, particularly those conducted prior to Hurlbert's (1984) definitive paper on experimental design, a single pair of disturbed and undisturbed (or one treated and one untreated) forest sites have been compared following treatment. While such studies have helped generate hypotheses to predict the effects of canopy-opening disturbances, herbivory, or defoliator suppression, a single pair of sites provides no error degrees of freedom for statistical analysis and does not permit the evaluation of confounding factors (e.g., the extent to which differences between the two plots are influenced by underlying edaphic or topographic factors such as gradients in soil conditions and atmospheric pollutants). For large-scale experiments, especially the assessment of disturbance or climate change effects, sufficient pretreatment data demonstrating substantial similarity among plots allows researchers to note any departures from these similarities following a treatment or environmental change (e.g., Davidson et al. 2008; Hurlbert 1984; Nepstad et al. 2007). Alternatively, combining multiple studies of such paired plots that used comparable sampling techniques provides sufficient independent replication and error degrees of freedom for the meta-analysis of treatment effects (e.g., Jactel and Brockerhoff 2007; Vehvilainen et al. 2007).

Experimental manipulation of canopy structure can be expensive and labor-intensive, as well as hazardous. The manipulation of canopy structure generally has been accomplished by cutting trees to alter canopy cover and/or composition—for example, thinning 13-hectare replicate plots to designated stem retention levels in aggregated or dispersed patterns to evaluate effects of canopy opening on forest structure and function (Fig. 4.3; Schowalter et al. 2005) or thinning overstory or understory and burning replicated 4-hectare plots in a factorial design to evaluate the effects of restoration treatments in mixed-conifer old growth on multiple forest structures and ecosystem variables (North et al. 2002; North et al. 2007). In one particularly ambitious project, climbers cut all canopy branches >10 centimeters in diameter in replicated 0.16-hectare plots in a tropical rain forest to simulate the effects of hurricane damage (Richardson et al. 2010; Shiels et al. 2010). The resulting debris was either left in the trimmed plot or relocated to a nontrimmed plot to permit the evaluation of the separate effects of canopy opening and the redistribution of canopy biomass to the forest floor on ecosystem variables.

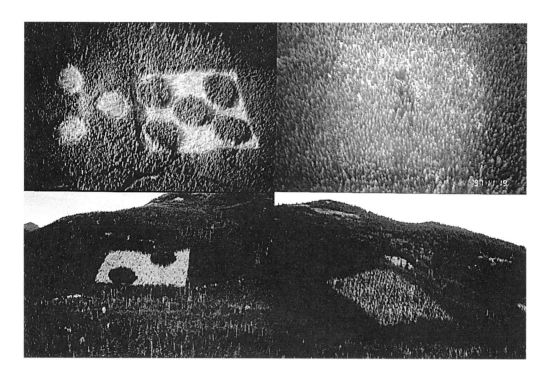

Figure 4.3.
Canopy retention treatments in the Demonstration of Ecosystem Management Options (DEMO)
experiment in western Oregon and Washington. Treatments clockwise from top left are: 75 percent
aggregated (left side of photo) and 40 percent aggregated (right side of photo) retention; 40 percent
dispersed retention; 15 percent dispersed retention; and 100 percent (right side of photo) and 15 percent
aggregated (left side of photo) retention.
Photos courtesy of the US Forest Service. (Top two photos by Jon Nakae and Jim White; bottom two
by Tom Savage.)

Manipulation of within-tree structure is relatively easier but also requires access and
attention to safety issues. Several studies have manipulated canopy structure in order
to evaluate effects on canopy biota. Whelan (2001) and De Souza and Martins (2005)
experimentally manipulated branch spacing and foliage density within tree crowns by
tying experimental branches to existing branches to evaluate the effects of these canopy
factors on bird and spider activity, respectively. Roubik (1993) experimentally manipulated the height of floral resources from different canopy strata in tropical forests in
Panama and demonstrated that the apparent fidelity of pollinator species to particular
canopy strata reflected pollinator preferences for particular floral resources. Most pollinator species were attracted to their preferred floral resources regardless of their location in the canopy. Lindo et al. (2008) placed artificial soil habitats of different sizes at
various heights within the canopy to assess the effect of soil volume, soil moisture, and
canopy height on the development of oribatid assemblages and found that assemblage

development reflected species tolerances to environmental conditions (e.g., fewer species in the harsh, moisture-limited upper canopy). Yanoviak (2001) manipulated predator and leaf-litter abundance in natural and artificial tree holes to evaluate effects on community development.

SUMMARY

Canopy structure refers to the framework of boles and branches and the distribution of foliage that represent the forest's engine for the capture and processing of energy and atmospheric nutrients, the template and resources for associated canopy flora and fauna (as well as the template of sample units for canopy research), and the mechanism by which forests modify local and regional climate, soil conditions, and associated stream ecosystems. Structural elements include vertical and horizontal variation in canopy surface topography, canopy architecture (branching pattern and angles), crown spacing, foliage distribution and density, reproductive structures, epiphytic mats, suspended soils, and aquatic habitats.

Measurement methods include the geometric representation of canopy structure and three-dimensional crown mapping using data derived from ground-based, within-canopy, or remote-sensing techniques. The selection of sample units depends on objectives. Forest canopies can be studied at multiple scales, from individual leaves or associated organisms to canopy communities and ecosystem processes at a global level. Smaller scales can be studied using traditional techniques (i.e., ground-based or within-crown methods). Larger scales require replicated units and are represented better by remote-sensing techniques. While much has been learned about forest canopies and associated communities of organisms using natural history approaches, many of our current questions require experimental testing of hypotheses, such as the effects of forest structure on global carbon flux and climate. Hypothesis testing requires manipulative methods that often are difficult in structurally complex forest ecosystems. However, manipulation of canopy structure has ranged from the suspension of experimental branches at various locations or angles, to test the effects of colonization by epiphytes or invertebrates, to the more ambitious cutting of all branches in large plots, to simulate hurricane disturbance. Harvest practices often provide replicated patches of modified forest that can be compared to test effects of canopy structure on gas exchange and local climate. Such studies will greatly improve our understanding and ability to predict the effects of modification or loss of forest canopies on biodiversity, carbon flux, and climate.

SUGGESTED READING

Dial, R., B. Bloodworth., A. Lee, P. Boyne, and J. Heys. 2004. The distribution of free space and its relation to canopy composition at six forest sites. *Forest Science* 50: 312–25.
Lefsky, M. A., A. T. Hudak, W. B. Cohen, and S. A. Acker. 2005. Geographic variability in lidar

predictions of forest stand structure in the Pacific Northwest. *Remote Sensing of Environment* 95: 532–48.

Sillett, S. C., and R. Van Pelt. 2007. Trunk reiteration promotes epiphytes and water storage in an old-growth redwood forest canopy. *Ecological Monographs* 77: 335–59.

Song, B., J. Chen, and J. Silbernagel. 2004. Three-dimensional canopy structure of an old-growth Douglas-fir forest. *Forest Science* 50: 376–86.

Treuhaft, R. N., G. P. Asner, B. E. Law, and S. Van Tuyl. 2002. Forest leaf area density profiles from the quantitative fusion of radar and hyperspectral data. *Journal of Geophysical Research* 107: 4568–80.

Turner, D. P., W. D. Ritts, W. B. Cohen, T. K. Maeirsperger, S. T. Gower, A. A. Kirschbaum, S. W. Running, et al. 2005. Site-level evaluation of satellite-based terrestrial gross primary production and net primary production monitoring. *Global Change Biology* 11: 666–84.

Van Pelt, R., and N. Nadkarni. 2004. Development of canopy structure in *Pseudotsuga menziesii* forests in the southern Washington cascades. *Forest Science* 50: 326–41.

CANOPY CONDITIONS, BIOTA, AND PROCESSES

Forest canopies present particular challenges with respect to sampling organisms. Collecting data on abundance or effects of organisms in forest canopies requires physical access, the ability to secure and monitor sampling equipment placed in the canopy, or remote-sensing capabilities. The difficulty of accessing canopy units for replicated sampling has been a major impediment to canopy study and has generated much of the creativity in canopy access methods described in Chapter 4. Taller or denser canopies require different sampling approaches than shorter or sparser canopies.

Sampling methods must be selected with respect to experimental objectives, safe and reliable access methods, and the particular organisms of interest (see Chapter 3). Methods appropriate for studying sessile plant variables cannot be used to study mobile canopy fauna (Lowman and Moffett 1994). Methods appropriate for studying canopy invertebrates are not useful for wide-ranging mammals or birds. Furthermore, the level of resolution must be appropriate for the research objectives. For example, studies of canopy invertebrates that identify specimens only to order, or even family, are not always adequate to evaluate food web relationships, since species within families and orders often represent multiple trophic levels (e.g., phytophagous versus detritivorous scarab beetles and phytophagous versus predaceous mirid bugs, both families common in forest canopies). In some cases, life-history stages of particular species require delineation because they have different trophic functions (e.g., herbivorous caterpillars versus flower-feeding/pollinating butterflies), complicating assignment of species to functional groups for biodiversity surveys (Schowalter 2011). One solution is to count collected caterpillars as herbivores and collected adults as pollinators or nonfeeders, as appropriate (see Heatwole 1989b).

This chapter describes sampling methods that have been used for studying canopy organisms and their responses to, and effects on, changing canopy conditions. Although most researchers will be familiar with the general methods appropriate to their fields, a description of the advantages and disadvantages of the various methods used in the canopy is intended to foster new approaches as well as collaborative and comparative projects.

OVERVIEW OF CANOPY BIOTIC VARIABLES

HABITAT CONDITIONS

Canopy environmental conditions determine the availability of habitat, food, and other resources for various organisms. Canopy conditions are affected by canopy structure and gradients in temperature, relative humidity, and airflow, as well as by distribution of nutrients in new and old foliage and accumulated litter and soil. These conditions establish food and habitat availability for associated organisms. In turn, epiphytes and parasites, herbivores, and detritivores alter canopy structure and distribution of soil and nutrients. Canopy conditions also change as a result of treefall and disturbances, such as fire, storms, and drought, as well as anthropogenic factors, such as climate change and atmospheric pollution. Rigorous methods for measuring these changes are necessary to document their effects.

Epiphytes represent a major component of the photosynthetic and water-holding capacity of the canopy (Díaz et al. 2010; Krömer et al. 2007; Obregon et al. 2011; Pypker et al. 2005; Sillett and Van Pelt 2007; Toledo-Aceves and Wolf 2008). Epiphytes also accumulate arboreal soil and litter and support development of the distinct communities associated with the arboreal soil/litter environment. Some epiphytes, such as the birdnest ferns of tropical forests, reach large sizes on crotches or large branches that have sufficient soil accumulation and are capable of supporting fern weights up to 200 kilograms fresh weight apiece (Ellwood et al. 2002). Increased weight of large epiphytes following heavy rains may cause breakage of smaller branches. Intermediate-sized epiphytes, such as most ferns, bromeliads, and orchids, occupy midcrown zones, and small epiphytes, such as lichens and mosses, occupy the outer smallest branches. These plants greatly increase habitat area for canopy fauna (Ellwood et al. 2002; Richardson et al. 2000) and for interception of airborne moisture and nutrients (see Chapter 6).

Plant parasites and endophytes are also important components of forest canopies (Carroll 1988; Shaw et al. 2005). Parasites include those growing externally, such as mistletoes, fungi, and strangler figs, as well as those growing internally, often indistinguishable from endophytes (Moffett 2000). Both groups affect canopy condition by removing nutrients, providing additional resources, and contributing chemicals that aid in defending the host plant (Carroll 1988).

Canopy animals are a diverse and important component of canopy communities. Invertebrates and birds are diverse and functionally important components of canopies in both temperate and tropical forests, whereas amphibians, reptiles, and mammals are more diverse and important in tropical forests. Invertebrates are particularly diverse in forest canopies (Basset et al. 2003; Basset et al. 2007; Erwin 1982; Gering et al. 2007; Novotny et al. 2002; Novotny et al. 2006). Although invertebrate diversity depends on tree species diversity and environmental conditions (e.g., Erwin 1982; Gering et al. 2007; Novotny et al. 2002; Novotny et al. 2006), most tree crowns host dozens to hundreds of species that represent specialized and generalized herbivores (including folivorous and sap-sucking species), detritivores, and generalized and specialized predators and parasites (Schowalter and Ganio 2003). The small size and heterothermy of these organisms make them particularly sensitive to vertical and horizontal gradients of temperature and relative humidity, as well as variation in resources in forest canopies. Many species, such as aphids, scale insects, and leaf miners, are small enough to live within individual leaves or within the boundary layer of plant surfaces that have relatively constant temperature and moisture conditions. Even smaller, tardigrades may be one of the more common groups inhabiting forest canopies (Box 5.1), but very few surveys have been undertaken (see Miller 2004). Canopy herbivores significantly affect foliage mass and canopy processes (reviewed in Lowman and Rinker 2004; Schowalter 2011. Epiphyte mats and accumulated soil host rich assemblages of arthropods

(Yanoviak et al. 2007). Arboreal collembola and oribatid mites contribute to litter decomposition within the canopy (Lindo and Winchester 2007). Other arthropods occupy the aquatic environments provided in tree holes (Yanoviak 1999) or canopy epiphytes (Richardson et al. 2000). Predators and detritivores occupy loose bark and crevices or rest on various canopy surfaces. Vertebrates include species that live and nest in forest canopies (such as tree frogs, birds, squirrels, and monkeys), as well as species that forage into or from tree canopies (e.g., snakes, hunting cats, and bats; Kays and Allison 2001; Malcolm 2004; Reagan and Waide 1995).

BOX 5.1: THE COLLECTION OF TARDIGRADES FROM THE CANOPY

William Miller and Margaret (Meg) Lowman

Tardigrades are a little-studied phylum of microscopic aquatic invertebrates found in the interstitial moisture that collects among the leaves of mosses and thalli of lichens that inhabit the trunks and branches of trees all over the world. Tardigrades are known for cryptobiosis, the process of desiccating as its habitat dries and reconstituting with the return of moisture (Miller 2004). Cryptobiotic tardigrades are dispersed on the winds (Kinchin 1994) and thus are rained onto the canopies of the world, where they must find acceptable habitat to survive (Miller 2004).

Despite our knowledge of tardigrade physiology, we know almost nothing about their ecology or basic requirements for living, nor do we know which species live in the canopy.

Tardigrade habitats (moss or lichen) can be found at all levels in the canopy. They include epiphytes and other surfaces including trunks, branches, and upper reaches of most species of trees. Thus the collection of the tardigrades is the collection of the habitat, with the extraction of the animals accomplished back in camp or the laboratory. Mitchell et al. (2009) suggest that tardigrades exhibit differential substrate selection and vertical distributional patterns.

The collection of tardigrade habitat in the canopy is a simple protocol executed while hanging from a rope high above the ground. Lush moss can be pinched off a limb, branch, or trunk by the handful, while thinner moss mats can be scrapped with a putty or pocket knife. Lichen generally needs to be scrapped with a knife, as do algae and biofilm. Sample selection should represent all the habitats found at each site (height). The standard collection container is a small brown bag (the type used to pack school lunches). Paper is preferred because the sample dries slowly, allowing the animals to desiccate naturally. But if it is particularly wet or damp, collections may be more practicable in small plastic sandwich bags and later transferred to the paper for more permanent storage.

Collection data are written directly on the bag with a felt-tipped pen. Core information includes the collector's name, date, collection code, name of location, name of site, latitude, longitude, substrate (tree species), altitude in tree, orientation (north, south, east, or west), girth of tree, habitat (moss, lichen, algae, etc.), and any other information deemed relevant. A GPS is required to locate the tree (latitude and longitude) in the forest, but the electronic altitude tends to be unreliable, so we recommend using a measured line or cloth tape on a reel to document the distance above the ground. A canvas shoulder bag with a zipper closure is suitable to carry the supplies (bags, pens, scrappers, etc.) and samples up and down.

Climbing equipment and rigging should be selected knowing that the purpose of the climb is the collection of the samples. Care must be given to the fact that at each sampling site it will be necessary to secure to the trunk or limb such that both hands are free to grab, pluck, or scrape habitat into the bag and then record data. A bosons seat or saddle and sling may be preferred equipment for both work and safety. It appears that this type of survey and collection would be ideal for researchers with ambulatory disabilities.

In the highest reaches of the tree, where the branches are thinner and security precarious, it is recommended to cut bunches of leaves with twigs and branches included. These larger samples can be folded or shredded to fit into a 1-gallon plastic bag.

In the lab or back at camp, transfer the collection data from the bag to your field notebook; then take a small part of the collected sample and place it into a cup or dish. A variety of containers will work, from ketchup cups to cereal bowls, as long as you standardize the habitat sample size; we use a 1 gram sample. Add 5 ounces of bottled water to the sample and let it soak for one to twenty-four hours. The drier the sample, a longer soaking time is necessary for the animals to become active, and the wetter the sample, the sooner searching is productive.

The larger samples are soaked in their plastic sample bags by adding water and allowing it to soak for a few hours. After soaking, the bag can be rolled on a table to break up the sample and free the animals into the water. Then the bag is shaken to float the animals, opened, and the water poured off quickly into a solid container (beaker, cylinder, or pitcher). Pour as fast as possible so the animals do not settle. A second rinsing with water under pressure can be flushed into the bag and poured off, as before. Let the container sit for a half hour so the animals settle to the bottom with the other debris. Using a larger siphon (50 ml syringe or turkey baster), suck up the bottom debris and concentrate it into a tube or dish. Let it settle, and subsample and examine, in the same fashion as the smaller sample.

From the samples, a disposable pipette is used to subsample the debris on the

(continued)

(continued)

bottom of the soaking dish and transfer it into a small, shallow search dish with a opaque (black) background. The dish is placed on a dissecting microscope at twenty fold magnification and illuminated with top-down reflected light. Twin fiber optics or LED lights seem to work best when one is placed on one side at a near ninety-degree angle to the dish and the other at a forty-five-degree angle. This results in good color and shadows in the sample, making animals easier to see and extract.

Tardigrades will be recognized as small (0.5 mm), caterpillar-shaped animals clinging to debris or floating free on the bottom of the dish (Box 5.1, Fig. 1). They may be white, reddish, orange, green, or transparent and may be actively moving about or asphyxiated and just lying on the bottom of the dish among the debris. A small micropipette or Irwin loop can be used to move them into a small tube of 70 percent ethyl alcohol or 4 percent buffered formalin. Tubes should be labeled with the collection code found on the bag and may be stored for extended periods.

For identification, tardigrades are put onto glass microscope slides in polyvinyl alcohol (PVA)-mounting media, which will partially clear the animal and make the claws and mouthparts more visible. Glass cover slips should also be used, as plastic is porous and discolors over time. Slides should be numbered to be traceable back to the data recorded on the collection bag. After a few days, the preparation is dry enough to be sealed with fingernail polish or epoxy paint to prevent long-term drying of the media.

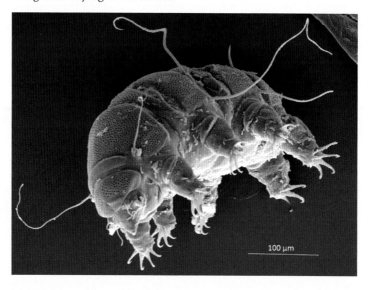

100 μm

BOX 5.1, FIGURE 1. Echiniscus trisetous, a common armored tardigrade in moss and lichen habitats found on trees.

Tardigrades are identified by the nature of their cuticle, the presence or absence of cirri, the size and shape of their claws, and the appearance of their buccal apparatus (Ramazzotti and Maucci 1983). More than a thousand species have been described to date, and it is expected that there are many species that are endemic to the canopy and just waiting to be discovered.

OVERVIEW OF CANOPY PROCESSES

Foliage production shows seasonal patterns in most forests, as a result of cold winters in temperate forests and dry seasons in tropical forests (Angulo-Sandoval et al. 2004; Brenes-Arguedas et al. 2008; Lowman 1992). Furthermore, individual leaves have lifetimes that range from a few weeks up to 20 years, depending on tree species and environmental conditions (Lowman 1999; Lowman 2009a). Leaf life span represents a trade-off between the cost of production and declining efficiency of photosynthesis. The cost of maintaining individual leaves may exceed their benefit when water or nutrients become limiting. Typically, limiting nutrients, especially nitrogen, are retrieved from foliage before abscission (Herrick and Thomas 2003; Marschner 1995). As a result, foliage abscised following normal senescence contains lower nutrient concentrations than fresh foliage falling as a result of disturbance or herbivory (Fonte and Schowalter 2004a).

Disturbance and herbivory increase the turnover of fresh foliage (as foliage fragments and prematurely abscised foliage) and rate of foliar nutrient flux (Chapman et al. 2006; Risley and Crossley 1988). Herbivory also increases the flux of nutrients via insect tissues and frass and the leaching of soluble nutrients from chewed foliar surfaces (throughfall; Fonte and Schowalter 2005; Kimmins 1972; Schowalter et al. 1991; Seastedt et al. 1983).

Many insects, birds, and bats are important pollinators of forest canopy plants. Bees commonly nest in forest canopies.

Although decomposition of foliage and woody material is best known from the forest floor, considerable detritus accumulates in forest canopies and either decomposes *in situ* or is preconditioned by canopy detritivores and decomposers before reaching the ground (Cardelús 2010; Coxson and Nadkarni 1995; Fonte and Schowalter 2004b; Paoletti et al. 1991). However, decay rates typically are slower in canopy litter than in forest floor litter (Cardelús 2010; Lindo and Winchester 2007). Detritivores, including collembolans and oribatid mites, are among the most abundant taxa in forest canopies (Schowalter 1995; Schowalter and Ganio 2003; Schowalter et al. 2005; Winchester 2006), suggesting abundant litter resources. Many termite species nest in forest canopies and contribute to the decomposition of woody and other detritus in the canopy.

Canopy litter varies in quality, depending on its source. Foliage has higher litter quality than wood, and tree species differ in litter quality (Fonte and Schowalter 2004a). Higher quality litter decomposes more quickly than lower quality litter. Carbohydrates and proteins are lost quickly, whereas cellulose and lignin decay more slowly.

Predation also is an important process in forest canopies. A variety of invertebrates and vertebrates prey on other canopy animals. Adequate predation rates can prevent herbivore populations from reaching levels that defoliate trees and alter canopy function. Conversely, relaxation of predation as a result of disturbance or other factors often triggers herbivore outbreaks.

MEASUREMENT OF ABIOTIC AND BIOTIC VARIABLES AND CANOPY PROCESSES

ABIOTIC CONDITIONS

Canopy microclimatic gradients can be measured by standard weather station devices placed at designated locations in the canopy (Madigosky 2004 Obregon et al. 2011; Parker 1995; Sillett and Van Pelt 2007). Light sensors, temperature/humidity probes, rain gauges, and anemometers, for example, can be located at various heights on towers or tree boles or positions between the bole and outer crown. Equipping sensor arrays with solar power sources facilitates continuous monitoring (Sillett and Van Pelt 2007).

Canopy leaf-area distribution and biomass and carbon, water, and nutrient content are measured by sampling tree and epiphyte foliage and accumulated litter and soil from various crown positions (Sillett and Van Pelt 2007). Samples can be collected from the canopy using the climbing techniques described previously and analyzed for water content using gravimetric techniques and for carbon and other nutrients using standard spectrophotometric techniques. Alternatively, remote-sensing techniques are now capable of providing detailed data for foliage area, biomass, and chemistry (especially water, lignin, nitrogen, and indirectly, phosphorus), using hyperspectral reflectance signatures, such as from airborne visible and infrared imaging spectrometry (AVIRIS; Chambers et al. 2007; Porder et al. 2005).

EPIPHYTES AND PARASITES

Adequate representation of canopy epiphytes and parasites requires stratification of canopy zones, ranging from large to small branches (Díaz et al. 2010; Krömer et al. 2007; Obregon et al. 2011; see Box 5.2). Larger epiphytes and external parasites can be surveyed from the ground using binoculars to estimate diversity and abundances per length of visible branch length (Ellwood et al. 2002; Shaw et al. 2005). Cranes, platforms, and tramlines within tree crowns can facilitate access to these zones for description of epiphyte and parasite composition and responses to changing canopy conditions (Shaw et al. 2005). However, direct access is necessary to quantify diversity or abundance of smaller epiphytes (e.g., mosses and lichens) and internal parasites or to sample epiphytes for biomass estimation (Díaz et al. 2010; Ellwood et al. 2002; Obregon et al. 2011). Manual removal and weighing of epiphytes is necessary to estimate biomass (Díaz et al. 2010).

BOX 5.2: SURVEYING EPIPHYTES IN FOREST CANOPIES

Cat Cardelús

It is challenging to take samples in the canopy of a tree. One must find safe, accessible trees to climb and often work under stressful conditions (e.g., the threat of insect stings, arboreal snakes, primates, and dangerous heights). Because sampling in the canopy is difficult, many individual canopy biologists have improvised novel sampling methods that work for them. The lack of standardization of these techniques makes cross-site comparisons difficult and limits our abilities to generate a synthetic understanding of the canopy system. Here, I break down epiphyte sampling methods into three tiers: short, intermediate, and long-term sampling. These categories are based on the major limitation to all research: time. In canopy sampling, this translates to finding host trees that are safe to climb and easily accessible. All three protocols produce excellent data but answer different questions about epiphytes, focusing on richness, richness and abundance, and diversity and distribution.

Many researchers work on a per-hectare basis, quantifying the number of epiphytes per hectare. I have found this limiting because a high density of accessible and safe trees are not often found in a randomly placed hectare. One way to deal with this is to quantify the number of large trees per hectare and extrapolate from the number of trees you measure in situ to the hectare scale. This can be problematic because not all large trees hold the same density of epiphytes, however.

SHORT-TERM SAMPLING

The short-term method is a rapid technique for the quantification of species richness. Often, researchers are interested in how many species of epiphytes are in a forest or region. Host trees vary significantly in their species richness. Such variation is linked to a number of diverse biotic and abiotic factors. Thus, maximizing host tree number, degree of branching, and distance on branch will result in more accurate richness estimates (Hietz and Ausserer 1998; Pittendrigh 1948). While epiphyte host specificity is rare, there are several studies that have detected host preference. For example, dry-season deciduous species tend to host more drought-tolerant epiphyte species such as tank bromeliads and cacti, compared to an evergreen host that may host more drought-intolerant species such as filmy ferns (Cardelús 2007; Pittendrigh 1948). Often more important than host-tree identity are the actual host-tree characteristics. Studies have shown that aspects such as branch diameter (Hietz and Ausserer 1998; Zimmerman and Olmstead 1992), branch height (Lyons et al. 2000; Zotz 1997), and bark water-holding capacity (Callaway et al. 2001) can significantly influence richness.

It is important to standardize the number of branches sampled as well as the area/branch sampled. Standardizing makes it possible to compare epiphyte species richness across trees

(continued)

(continued)

within your study as well as across studies. Species counts (presence/absence) per branch can be analyzed using sample-based rarefaction curves—species accumulation curves as a function of occurrence with estimates (Colwell 2005)—to determine if species saturation is reached (e.g., Box 5.2, Fig. 1; Cardelús et al. 2006; Watkins et al. 2006).

INTERMEDIATE SAMPLING

The intermediate sampling method samples for both species richness and abundance (diversity) as well as distribution. This method requires more data collection within the canopy. You must record not only the number of epiphyte species but also the abundance of each epiphyte along branch transects or within plots on the branch, as well as detailed information on the distance from the trunk and distance from the center of the plot (Cardelús 2007). These data sets are rich to analyze and informative for community-based analyses. These data lend themselves to nonparametric analyses such as discriminant function analysis and/or canonical correspondence analysis, as well as neighborhood analysis (Cardelús and Chazdon 2005; Cardelús 2007).

Maximizing host tree replication is essential for a community-based epiphyte study. Too often researchers thoroughly sample only one individual tree. While these data are informative and interesting, single-tree sampling is problematic because no statistics can be used to analyze the data—an N = 1 only characterizes that tree. However, if you sample two to four branches for each of three individuals of the same tree species, you have the statistical power to generalize about tree species.

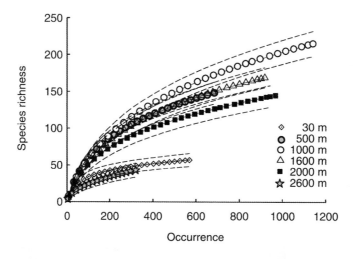

BOX 5.2, FIGURE 1. Sample-based rarefaction curves and 95% confidence intervals for observed vascular epiphytes at each elevational study site along the Barva Transect elevational gradient, Costa Rica. (From Cardelús, et al. 2006, Figure 3)

LONG-TERM SAMPLING

The long-term sampling protocol incorporates species richness, abundance, and distribution, as well as demography. There are very few studies that have measured the growth rates and population dynamics of epiphytes over time (Watkins et al. 2007; Zotz 2005). These studies are more labor intensive because they require returning to the epiphyte populations. They are enhanced by measuring plant functional traits such as plant height, clonal plant production, leaf life span, specific leaf weight, leaf size, and leaf nutrient concentrations (Cornelissen et al. 2003). Permanently tagging leaves with aluminum tags or trimmed plastic binding combs allows for easy return to the plant.

BOX 5.2, TABLE 1.   List of Variables to Measure and Their Methods for Each Protocol: Richness, Diversity, and Demography*

Variable	Methods	Species Protocol
Host tree diameter (cm)	With DBH tape, measure the diameter at breast height.	Richness
Branch height (m)	Use tape measure or laser finder. In dense forest, it is easiest to tie the tape measure down at the base of the tree and climb with the tape measure. If not a dense forest, a digital laser finder works well.	Richness
Branch diameter (cm)	With DBH tape, at the base of the branch, measure its diameter.	Richness
Branch angle	Use a protractor to measure angle.	Richness
Branch azimuth (degrees)	With a compass, measure the branch-facing direction in order to determine when the light reaches it.	Richness
Canopy soil depth (cm)	With a caliper that has a metal bar that comes out the base, dip the metal bar into canopy soil perpendicular to the branch until the base is flush with the top of the soil.	Richness

(continued)

Variable	Methods	Species Protocol
Species number	Record each species or morpho name of each species on the branch or within the study area. Collect at least one individual of each species for identification.	Richness
Number of individuals/species	Count the number of individuals of each species per plot/transect/branch.	Diversity
Distance of individual from trunk (cm)	With a tape measure, measure the distance from the base of the branch to the epiphyte.	Diversity
Distance of individual from edge of branch (cm)	With a tape measure, and noting the side from where you measure, measure the distance to the epiphyte. This will orient the epiphyte on the branch to examine neighborhood effects.	Diversity
Leaf demography	Mark each leaf with a permanent paint pen or leaf tag. Change the color of pen or tag for each census in order to track leaf longevity.	Demography
Epiphyte height (cm)	With a tape measure, measure the height of the epiphyte from the base of the plant.	Demography
Epiphyte specific leaf area (SLA; $g \times cm^{-2}$) or leaf mass per unit area (LMA; $cm^{-2} \times g$)	Take three to five punches of foliar tissue with a hole punch with known area (cm^2). Measure the dry weight of the hole punches and take the mean. This can be presented as SLA or LMA.	Demography, optional

Variable	Methods	Species Protocol
Epiphyte stoichiometry (percent N, percent C, percent P)	This requires a large foliar sample that is dried, ground, and rolled for elemental analysis (percent N, percent C) or ash digested and colorimetry for percent P.	Demography, optional
Host tree specific leaf area (g × cm⁻²) or leaf mass per unit area (cm⁻² × g)	As per epiphytes, this requires three to five punches of full-sun foliar tissue collected from the top of the canopy. Collection can be made with a shotgun or a crossbow.	Demography, optional
Host tree stoichiometry (percent N, percent C, percent P)	As for epiphytes, this requires foliar samples that are dried, ground, and rolled for elemental analysis (percent N, percent C) or ash digested and colorimetry for percent P.	Demography, optional
Canopy soil stoichiometry (percent N, percent C, percent P)	Volumetrically sample the soil (measure the area and depth of sample), dry, grind, and roll for elemental analysis (percent N, percent C) or ash digested and colorimetry for percent P.	Demography, optional

* *Diversity* includes the methods for richness, while *demography* includes methods for *diversity*.

INVERTEBRATES

A variety of methods have been developed to measure invertebrate abundances or diversity and their effects on canopy structure and processes (Fig. 5.1), but techniques vary in their level of accuracy (Lowman et al. 1996; Ozanne 2005) and representation of invertebrate taxa (Basset et al. 2003, 2007; Blanton 1990; Leather 2005; Majer and Recher 1988; Southwood 1978). Therefore, a method appropriate for the group of interest, or a combination of techniques to represent a broader diversity of groups, should be selected. Sampling techniques for invertebrates also differ in their ability to provide qualitative or quantitative data. Qualitative methods provide primarily diversity and relative abundance data. They do not represent specific units of area that can be used to calculate population density. Common methods include branch beating and interception traps. Quantitative sampling methods represent units of area that permit calculation of population density for comparison among experimental treatments or forest types. Quantitative methods include branch bagging, epiphyte mat sampling, and canopy fogging.

Branch beating involves beating a foliated branch over a collecting sheet, using a heavy stick or bat. Dislodged insects are collected from the sheet (Lowman 1985; Southwood 1978). An aspirator is useful for collecting small or fast-moving arthropods. Branch beating is most practical where the canopy-access method leaves both hands free to hold the bat and sheet. This method can provide a somewhat quantitative estimate of population density if the beaten foliage area or branch length is measured.

Various types of interception traps have been developed for sampling invertebrates. Malaise traps, flight interception traps, and light traps are the most commonly used (Fig. 5.1A, B, C). These traps can be suspended from ropes or attached to bole or branches at different heights to represent vertical distribution of arthropods (Hirao et al. 2007). An advantage of interception traps is that they collect passively over a designated time period (usually five to fifteen days), thus representing a substantial portion of the diversity of insects, flying both day and night. A disadvantage is that these traps do not represent nonflying arthropods or sessile and internal (galling or mining) arthropods, which often are the most abundant and functionally important members of the canopy community (e.g., Schowalter and Ganio 1998; Schowalter and Ganio 2003; Fig. 5.1A, B, C).

Malaise traps are large mesh tents that funnel a variety of flying insects into a collecting jar at the peak of the tent. The size of these tents makes positioning in the canopy problematic, and they are easily damaged by high winds, but large numbers of invertebrates, especially Lepidoptera, Diptera, and Hymenoptera, can be collected using this technique (Grimbacher and Stork 2007; Southwood 1978).

Flight interception traps are constructed of a fabric, glass, or clear plastic panel suspended over collecting pans filled with preservative (Southwood 1978). These traps exploit the tendency for many flying Coleoptera and some other insect groups to fall when they hit an obstacle. The diversity of insects collected can be increased by coating the panel with adhesive to retain all insects landing on the surface. However, the quality of specimens (especially of flies and moths) for identification may be reduced by damage from the adhesive. Attention must be paid to the weight of the preservative-filled pans, the rate of evaporation, and the toxicity of the preservative, which may attract thirsty animals. Propylene glycol (animal-safe antifreeze) is preferred as a preservative. Traps generally are checked every five to fifteen days, depending on the evaporation rate.

Light traps consist of an ultraviolet or fluorescent light placed against a white sheet

FIGURE 5.1.
Methods for sampling invertebrates: Malaise traps (A) near the ground, showing detail of funnel structure and collecting container at apex, (B) suspended midcanopy, and (C) at the top of the canopy; (D) branch bagging using a long-handled insect net from the ground *(continued on next page)*.

or panel that funnels attracted insects into a collecting jar (Lowman 1985; Southwood 1978). Nocturnal flying insects, attracted to the light source, land on the sheet or fall into the jar and can be collected from either the sheet or the jar (Hirao et al. 2007). Lights can be powered by portable generator or battery, requiring attention to support these heavy items and/or electric cabling in the canopy.

Branch bagging (Fig. 5.1D, E, F) involves slipping a closeable plastic bag over a foliage-bearing branch to collect invertebrates on the branch (Blanton 1990; Majer and Recher 1988; Schowalter and Ganio 2003). Accessible branches can be bagged from towers or crane gondola, by climbing into the canopy, or from the ground using a long-handled net. Samples can be collected at measured heights, most easily from towers or cranes, to evaluate the effect of crown height and environmental conditions on invertebrates (Schowalter and Ganio 1998). Long-handled nets can reach branch ends at heights of 10 to 15 meters, depending on slope. A drawstring is inserted around the rim of a plastic bag, the open end of the bag is attached to the rim of a net at the end of a telescoping or extendable pole, the net is slipped quickly over a foliage-bearing branch, the bag is closed with the drawstring, and the branch is clipped for sorting of invertebrates in the lab. This technique represents abundances of nonflying scale insects, gall formers, and leaf miners that are unsampled by other techniques, and it provides quantitative estimates (number of insects per unit leaf area or mass) of abundances of invertebrate species most characteristic of the particular tree species sampled. This technique has maximum mobility among trees, but elevating the pole through dense understories may limit use

Figure 5.1. *(continued)*
Methods for sampling invertebrates: (E) using rope access, or (F) using a canopy crane *(continued on next page)*.

in some forests, and only accessible branch ends can be sampled. Furthermore, this technique provides only a snapshot of invertebrates in the sample and likely underestimates the abundances of highly mobile (especially flying) invertebrates. Nocturnal sampling is necessary to represent nocturnal invertebrates (Schowalter and Ganio 2003). Foliage in samples also can be used to estimate the effect of herbivory (discussed later; Fig. 5.1D, E, F).

Canopy fogging (Fig. 5.1G) involves extending a hose into the crown from a sprayer that blows an insecticidal mist into the crown (Blanton 1990; Dial et al. 2006; Erwin 1982; Floren 2010; Floren and Schmidl 2008; Gering et al. 2007; Majer and Recher 1988). Unattached invertebrates fall from the crown and can be collected on sheets

FIGURE 5.1. *(continued)*
Methods for sampling invertebrates: (G) canopy fogging.
Photo G courtesy of Andreas Floren.

stretched under the crown. Site proximity to a road is necessary to facilitate access for fogging equipment. This method only provides a quantitative measure of invertebrate abundance if numbers of specimens are divided by the area of the collecting surface. This technique represents a large portion of overall diversity but does not represent scale insects, gall formers, leaf miners, or other invertebrates incapable of falling to the ground (e.g., small invertebrates intercepted by dense conifer needles) or insects capable of escape.

More focused sampling of target groups may involve the use of attractants to lure relatively dispersed individuals to baited traps. For example, Roubik (1993) used traps baited with cineole to compare abundances of euglossine bees at different canopy heights in a tropical rain forest using a canopy crane. Heatwole (1989a) used various attractants to measure abundance and distribution of ant species along a forest gradient. Malaise traps baited with ripe or rotting fruit can be used to attract butterflies or flies (e.g., Bossart et al. 2006; Fig. 5.1A–C).

Arthropods associated with perched soil or epiphyte habitats can be extracted from these habitats for measurement using Berlese funnels. This method involves placing a sample in a funnel suspended under a light/heat source and collecting invertebrates moving away from the adverse conditions in a collector placed under the funnel (Lindo and Winchester 2007).

Sampling methods differ in their representation of various invertebrate taxa, as noted previously. Several studies have compared sampling methods in terms of representation of canopy invertebrate diversity and abundance. Blanton (1990), Majer and Recher (1988), and Ozanne (2005) compared branch bagging and fogging methods and concluded that branch bagging provided larger numbers but fogging represented greater diversity. Depending on the degree of complementarity (i.e., the overlap of species representation) of sampling methods, combinations of methods can improve efficiency of canopy invertebrate sampling. For example, Longino and Colwell (1997) reported that Berlese funnels, Malaise traps, and canopy fogging individually were equally efficient in capturing ants but differed in the types of ant species collected. Combining Berlese samples with either Malaise or fogging methods improved the efficiency of representation of total ant diversity, compared to individual methods, because the species sampled were highly complementary. Combining Malaise and fogging methods did not improve efficiency because of the high overlap (i.e., low complementarity) of the faunas collected.

VERTEBRATES

Vertebrate abundances and diversity can be measured using either qualitative or quantitative methods, as described previously. Qualitative measures (such as visual or acoustic surveys, camera traps, mist nets, or dung surveys) have been used to confirm presence and relative abundances among species. Bait stations or feeders can be used with many

BOX 5.3, FIGURE 1. (A) Canopy traps affixed in the treetops, which photographed (B) macaques from Western Ghats, India.

BOX 5.3, TABLE 1. Some Key Factors That Can Influence Camera Trapping in the Canopies and Possible Solutions Toward Combating It

Interference	Factor	Effect	Solution	Remarks
Weather	Rain	Water	Provide multiple covering, but make sure this is not left alone for long, as internal condensation might occur.	
	Mist	Image blurred	A transparent thin oil is used sometimes to let water droplets drain out.	
	Cold	Battery power	Frequency of battery change needs to be increased, or supplement battery with an external car/bike battery.	
	Sun	False trigger	Face north-south, with camera facing down.	A combination of sun and wind can completely drain the battery or film through false trigger within few hours.
	Wind	False trigger	Attach on the main trunk, if possible.	
Animals	Macaques	Can dismantle	Have a protection box that does not allow them a grip.	
	Squirrels	Can chew the cover	Use protection box with metal anchor wires.	

Interference	Factor	Effect	Solution	Remarks
	Ants	Can get inside and corrode circuits	Close all openings.	
Power		Rechargeable cells can drain fast	Check every 2nd day	Solar power can be an option but there is an issue of protecting this from monkeys etc.
		Long lasting alkaline cells can drain	Check every 4-5 days	
Human		Can be vandalized if noticed	Camouflage	

Based on our experience, we would recommend the following for successful camera trapping in the canopy:

1. Choose the tallest tree whenever possible and fix the camera at the subcanopy level. This gives you a fair advantage to point the camera below the horizon and in north-south direction to prevent sun–wind interference. Be sure it does not look at the canopy of the other tree and effectively points to a dark area in the understory.

2. Choose a thick branch, if for only diurnal monitoring. Nocturnal monitoring can be done by fixing the camera on other branches, which may sway in the wind since wind alone will not trigger the camera.

3. Use a synchronized slave flash to increase the reach of the flash, since the infrared sensors work for about 60 to 100 feet but the flash will not.

4. Do not plan camera trapping on rainy and windy days. If essential, protect the housing by camouflage polythene, and check the cameras every 24 hours (or, even better, 12 hours).

5. Use alkaline cells, as these last longer than rechargeable cells. In any case, if there is evidence of higher usage of power by the camera, you need to check the camera every day.

6. Spray ecofriendly degradable repellent on the trap if animals are a problem (although we haven't tested the repellent's effectiveness).

7. In addition, train in canopy access before climbing.

(*continued*)

(continued)

Camera trapping in the canopy is not very effective when the object is partially hidden by foliage like fruits and flowers that don't pop up. It was also not very effective in capturing bats. We would, however, recommend video camera traps; these are extremely useful in such cases. There are issues of power, cost, and memory space, but these can be easily overcome with long-lasting batteries and external power sources such as car or motorcycle batteries. With memory-card space exploding and camera prices falling, this should also not be a serious issue.

Lastly, camera trapping is an excellent tool for capturing rare and elusive animals, but nothing exceeds the pleasure of sitting for long hours under a fruiting or flowering tree, watching and listening to the life around you.

Mist nets can be positioned in the canopy to collect flying birds or bats. This technique is most convenient when the net can be stretched across flyways between tree crowns, especially if accessible from towers, but raising nets and preventing damage to net or vegetation is difficult (Munn and Loiselle 1995). This technique also can be used to collect birds and bats for mark-recapture studies (as discussed later). Wunderle et al. (2006) mist netted birds to compare diversity and assemblage structure between uncut and reduced-impact harvests in Amazonian rain forests. Stouffer et al. (2009) used mist netting to compare extinction rates of birds in Amazonian forest fragments of varying sizes.

Live trapping provides information on occurrence and habitat preferences for small mammals. For example, Meyer et al. (2007) attached live, baited traps to boles of old-growth conifer trees to compare frequencies of flying squirrel captures among sites varying in canopy density, leaf litter depth, and proximity to streams.

Finally, many vertebrates can be surveyed by locating their dung (scats), especially on the ground under nests or foraging trails through the canopy. Julliot (1997) surveyed red howler monkey defecation sites under sleeping sites in trees to evaluate the monkeys' role in seed dispersal.

Quantitative sampling methods include mark-recapture trapping for the measurement of foraging or territory sizes. This method requires considerable time and labor to set up mist nets or a network of traps (in three-dimensional canopy space) to collect individuals that are then marked in various ways (e.g., florescent dyes or ear tags) for future identification when recaptured (Kays and Allison 2001; Malcolm 2004; Reagan 1996; Stewart and Woolbright 1996). However, to be most effective, traps must be positioned in ways most likely to capture target animals, and the marking technique must avoid injuring the animal or making it more obvious to predators—either of which would reduce the likelihood of recapture and compromise results. The pattern of recaptures among traps in the network can be used to measure the area of individual foraging territories, thereby providing an estimate of population density.

Several methods have been employed to measure foliage turnover and herbivory (Boxes 5.4 and 5.5). The most informative method is the frequent monitoring of marked leaves (Lowman 1984; Lowman 1992). This method permits the measurement of leaf consumption as it occurs. As leaves expand, holes chewed by herbivores also expand, so that direct area (cm² or mm²) measurement after leaves have expanded generally leads to overestimation of herbivory, although the percentage missing (an estimate of herbivore effect on canopy cover) stays the same (Lowman 1984). Life table analyses applied to the fate of marked leaves permits assessment of the importance of factors causing foliage loss. However, this method requires frequent access to marked leaves, most conveniently provided by a tower or crane, and is labor intensive.

The periodic measurement of the leaf area missing due to herbivory is probably the most widely used method for estimating herbivory. This method does not represent herbivory per se, and provides only a snapshot of leaf area missing at a point in time, but does indicate the effect of herbivory on leaf area index and canopy function (Schowalter and Lowman 1999). Measurement of leaf area missing generally is done with a leaf area meter on leaves collected from towers or cranes or in bagged branch samples (e.g., Hargrove 1988; Hargrove et al. 1984; Lowman 1984; Schowalter and Ganio 2003). Leaves must be pressed flat for measurement in a leaf area meter or copied onto paper while fresh, which avoids shrinkage during drying and pressing. In either case, area missing is measured as the difference between leaf area with holes and leaf area with holes covered. Real leaves can have holes taped and leaf margins reconstructed. Copied leaves can be measured before and after cutting out holes.

Herbivory can be measured directly by placing mesh sleeve cages over herbivore-bearing branches and monitoring the rate of consumption and foliage loss (e.g., Connelly and Schowalter 1991; Fonte and Schowalter 2005). Sexton and Schowalter (1991) used aluminum baffles around the lower boles of trees to restrict access to a flightless weevil that feeds on Douglas-fir conelets and foliage for evaluation of conelet losses with weevils present or excluded. Various exclusion methods also can be used to measure herbivory in the presence or absence of predation (as discussed later).

Herbivory can be estimated by the rate of litter or frass collection in litterfall traps on the forest floor (Reynolds and Hunter 2001) and from the rate of foliage consumption by individual herbivores, scaled up to population densities in the canopy (Stadler et al. 1998), based on estimates of consumption by herbivores of 50–150 percent of their dry body mass per day (Reichle and Crossley 1967; Reichle et al. 1973). However, frass disintegrates easily when wet, so it must be collected before precipitation events. Mizutani and Hijii (2001) measured the effect of precipitation on frass collection in conifer and deciduous forests and calculated correction factors to improve the estimation of frassfall. Donati et al. (2007) monitored the feeding activity of collared lemurs in Madagascar by observing fruit, flower, and leaf selection during the day or identifying fragments falling from feeding perches at night.

Remote-sensing techniques have become sufficiently sophisticated to measure effects of herbivory, as well as various plant stressors, on a variety of coniferous and deciduous trees (Carter and Knapp 2001). Tree stress is expressed as increased reflectance at wavelengths near 700 nanometers. This optical response reflects a general tendency for stressed foliage to reduce chlorophyll concentrations. Nansen et al. (2009) reported that the experimental infestation of wheat stem sawfly, *Cephus cinctus*, on wheat was detectable as significantly reduced reflectance at 725 nanometers. The normalized difference vegetation index (NDVI) and photochemical reflectance index (PRI) decreased in response to sawfly infestation, whereas the stress index (SI) increased. Nansen et al. (2010) evaluated effects of severe, moderate, or no drought stress or spider mite (*Tetranychus urticae*) infestation in cereal crops and found a particularly strong response to drought (but not spider mite) stress at 706 nanometers and a significant response to spider mites, as well as drought stress, at 440 nanometers. Remote sensing has not been used yet to measure herbivory in forest canopies but offers a promising new approach to such measurement.

DECOMPOSITION

Decomposition has been measured using various methods (Schowalter 2011). However, the most convenient method for use in forest canopies is tethered litter or litterbags.

Multiple weighed samples of leaf litter or small woody fragments can be tethered to branches at each of several locations in the canopy, chosen to represent different canopy environmental conditions. At designated times, one sample from each location can be collected for reweighing to measure mass loss per unit of time. This method is vulnerable to losses due to undecomposed fragments falling from the canopy.

Litterbags partially solve the problem of loss of undecomposed mass. Weighed leaf or woody litter is enclosed in mesh bags that are then tethered to branches in the canopy, as noted previously (Cardelús 2010; Lindo and Winchester 2007). Selective repellents or toxins (e.g., naphthalene) or different mesh sizes can be used to restrict access to litter in order to assess the relative contributions of different canopy organisms to decomposition processes (Ingham 1985). At designated time periods, scheduled to represent fast (higher quality) and slow (lower quality) components of litter decomposition, a subset of litterbags are removed, dried, and weighed to measure mass remaining at that time.

Regardless of method, decomposition rate is calculated as a single- or double-component process (Olson 1963):

$$N_t = S_o e^{-kt} + L_o e^{-kt},$$

where N_t is mass at time t; S_o and L_o are masses in short- and long-term components, respectively; and k's are the respective decay constants.

BOX 5.4: BEETLES IN A SALAD BAR: ACCURATE ASSESSMENT OF HERBIVORY IN WHOLE FORESTS

Margaret (Meg) Lowman

This book had its genesis in discussions about the inaccuracy of many methods of assessing herbivory in forest canopies. A quick review of the literature reveals that most forest–herbivory studies were limited to understory leaves collected at one point in time, akin to taking a snapshot of one city block and assuming that it is typical of the entire urban landscape and, even worse, that it will remain the same over its life span. Leading up to this book, these conversations usually occurred in the canopy of some remote rain forest, reinforcing the notion that many aspects of forest canopy science might benefit from defining current methods for other aspects of canopy research, not just herbivory. It became apparent to the community of canopy researchers that a methods book would encourage "best practices" in studies of whole forests, even though such a book is limited by its publication date as a "snapshot" of sorts, since scientific methods continue to improve.

Historically, most studies of herbivory have been conducted in an overly simplified fashion. First, ecologists often overlook the fact that there are many types of herbivory—foliage feeding, sap sucking, bark boring, galling, and leaf mining, to name but a few. It is essential to define the types of damage (for any ecological measurements) caused by removal of plant tissue up in the "salad bar in the sky." Second, many field studies rely on discrete samples of leaves collected at one point in time, which represents a snapshot of one season and potentially of one age class of leaf materials. Third, most foliage studies focus on the understory, usually a narrow band of foliage within arm's reach from ground level to 2 meters high, which represents less than 10 percent of the whole forest. With the knowledge that most insects, especially beetles and other major herbivores, reside in the upper canopy, the understory leaves are not likely to be representative of the whole forest. Not surprisingly, herbivory levels in earlier literature that relied only on these overly simplified methods of sampling understory all tended to average 5–8 percent of leaf area missing and attributed to herbivory (e.g., Bray and Gorham 1964; Landsberg and Ohmart 1989; but see Lowman 1984). Sampling in this fashion is akin to having a doctor examine your thumb and then provide a prognosis of your entire body health. Approximately 95 percent of the forest remains unstudied in cases where a small portion of the understory is extrapolated into whole-forest herbivory assessments. And even less information is revealed when herbivory is measured discretely or as a snapshot, rather than over the life span of the leaves.

To overcome these shortcomings, the advent of experimental ecology has led to more rigorous and accurate sampling within forests. Herbivory is no exception. The traditional techniques of measuring leaf area losses to foliage feeders by destructive sampling of small quantities of understory leaves is now considered inadequate for quantifying whole-forest

(continued)

(continued)

processes, especially since the upper canopy is known to be the "hotspot" for both insect herbivores and foliage growth (Lowman 1984). Sampling herbivory requires careful attention to both temporal and spatial factors, with replication of those variables considered important for the hypothesis addressed. In the case of lowland tropical rain forests with diverse species and long-lived evergreen foliage, important temporal variables include day and month (as young leaves are eaten significantly more than older leaves, with most leaf consumption occurring in young leaves; Lowman 1985), year (with respect to both outbreaks and climate change, which may be decadal events), and ages of leaves (since some rain forest leaves remain on the tree for more than twenty years; Lowman 1995; Lowman 1999). With respect to spatial factors, herbivory varies among leaves, branches, height above ground, individual crowns, and different forest stands and between types of forest (Box 5.4, Fig. 1). Depending on the hypothesis addressed, sampling may require replication with respect to all these variables.

When herbivory is measured with respect to spatial and temporal factors, the results are more accurate. When leaves are monitored over time throughout their life spans—also called *long-term sampling* (as compared to *discrete sampling*, which is a snapshot assessment)—the results are most accurate. In the case of *Dendrocnide excelsa* in Australian subtropical rain forests, herbivory was measured as less than 2 percent if sampled at the onset of leaf emergence (Box 5.4, Fig. 2A) or as high as 42 percent when the leaves were mature and a host-specific herbivore called *Hoplostines viridipennis* has found its food plant (Lowman 1985; Box 5.4, Fig. 2B). Long-term

(continued)

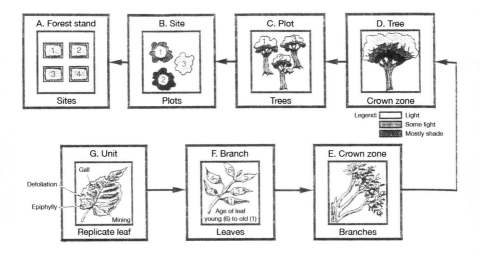

BOX 5.4, FIGURE 1. Experimental depiction of sampling design to measure herbivory throughout forest canopies, scaling up from leaf to branches to crown zone (height, light) to individual trees to site to forest stands.

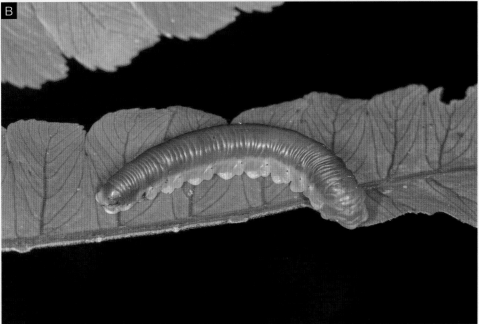

BOX 5.4, FIGURE 2. Examples of herbivores feeding on foliage in the canopies of Amazon lowland forests, whereby some larvae feed gregariously (A) and others in isolation (B). Photos courtesy of Phil Wittman.

(continued)

sampling (i.e., monitoring leaf damage *in situ* every month) is the only way to accurately quantify leaves that may be totally (or significantly) eaten by herbivores, since they do not show up in discrete or snapshot assessments, other than a small petiole sometimes attached to a branch.

With more than two decades of herbivory assessment in seven major forests around the world, I compared discrete (or snapshot) herbivory assessments with long-term measurements of herbivory. The difference was astounding. Overall, discrete measurements underestimated herbivory by 2.5 fold. In other words, for every intact leaf in a canopy, approximately 2.5 leaves were totally consumed. With canopy access, the ability to measure whole-forest herbivory—thereby reaching the upper canopy where entire leaves are frequently consumed—

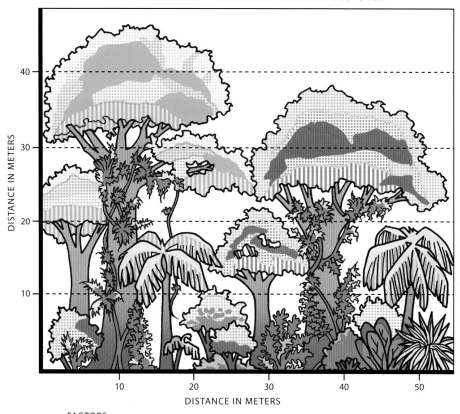

HETEROGENEITY OF FOLIAGE IN A FOREST ECOSYSTEM

DISTANCE IN METERS

DISTANCE IN METERS

FACTORS:
Height, Light, Species, Individuals, Age (Leaves, Plants) Site, Density, Diversity, Human Activity

BOX 5.4, FIGURE 3. Map of a forest canopy, illustrating the spatial and temporal variation in leaf quality and indicating the need to replicate herbivory measurements throughout the whole forest over time.

has led to a correction: herbivory is much higher than previously measured by limited snapshot, understory sampling (Lowman 1984). In one carefully measured forest site of Australian subtropical rain forest, the canopy was more heterogeneous with respect to height and light of individual crowns than between species of trees (Box 5.4, Fig. 3). For ecological monitoring, this means that canopy access, replication with respect to temporal and spatial factors, and long-term observations are essential to quantify the process of herbivory. And for such in situ measurements over the life span of a leaf, additional time-consuming measurements of permanently marked leaves are most accurate: tracing leaves or using photographs that can be assessed by an imaging process back in the lab are more accurate than casual estimates of percentage leaf area missing (reviewed in Lowman 1984). With the increasing episodes of insect outbreaks as climates become warmer and drier, accurate methods to quantify insect herbivory in whole forests are critical to our understanding of forest health (Lowman 2009a).

BOX 5.5: MATHEMATICAL MODELING OF FOREST CANOPIES FOR HERBIVORY

Leon Kaganovskiy and Margaret (Meg) Lowman

Tree data structures can be used to model the dynamics of tree canopies attacked by insects, based on cellular automata or percolation ideas similar to fire and infectious disease propagation models. We developed a stochastic model of a generic tree where, at each step, probabilistic distributions are used to predict if a particular tree branch will divide into subbranches (number of branches as well as angles are probabilistic as well) or finally terminate with leaves. Once such a tree structure is created, a volume containing leaves is determined and broken into a large number of computational cells. Then the dynamics of the foliage including herbivory, aging, and other temporal and spatial factors is first evaluated within the computational cells and mapped back onto the tree crowns.

We use the following color coding for the five states of leaves: light green for young leaves, green for mature leaves, light red for slightly eaten leaves, red for largely eaten leaves, and black for dead leaves. If any of the twenty-six three-dimensional neighbors of a given leaf is attacked by bugs, we allow for the possibility of the spreading of insects to the given leaf with user-defined probability. The computational results show trends similar to extensive field measurements (Box 5.5, Fig. 1).

In addition, we considered a continuous model in which a certain small percentage of a leaf is eaten daily, which provided good comparison with recent daily measurements performed by New College of Florida students. Thus, this percolation/cellular automata approach can produce realistic descriptions of tree canopies being attacked by insects; the predictions correlate well with real-life experiments.

(continued)

(continued)

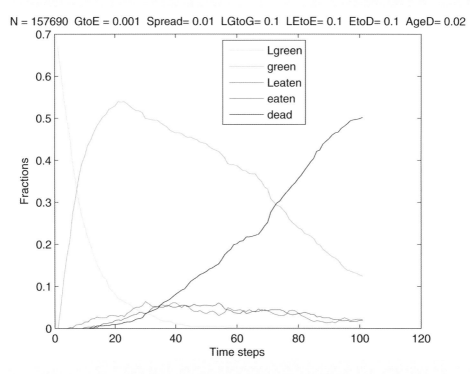

N = 157690 GtoE = 0.001 Spread= 0.01 LGtoG= 0.1 LEtoE= 0.1 EtoD= 0.1 AgeD= 0.02

BOX 5.5, FIGURE 1. This figure shows how the fractions of each type of leaves damaged by herbivory behave over time in our model.

On the top of the figure we list parameters of the this particular run. The variables are as follows:

N - number of points, p_GtoE - probability of a leaf being attacked by herbivores;

p_LGtoG - probability of going from light-green to green;

p_LEtoE - probability of going from light-eaten (light-red) to eaten (red);

p_EtoD - probability of going from eaten (red) to dead;

p_AgeD - probability of a leaf dying due to age.

PREDATION

Predation has rarely been measured in forest canopies, although the density of canopy predators alone would indicate high rates of predation (Schowalter 1995; Schowalter and Ganio 2003). Predation often maintains trophic cascades that limit canopy herbivory (Dyer and Letoureau 1999a, b; Mooney 2007; Terborgh et al. 2001).

A comparison of prey densities or population growth when predators are present or absent provides an estimate of predation rate. Terborgh et al. (2001) compared abundances of herbivores (howler monkeys, rodents, iguanas, and leaf-cutter ants) between mainland sites (with vertebrate predators) and islands, created by a hydroelectric impoundment in Venezuela, that were free of vertebrate predators. The herbivorous species all were ten to a hundred times more abundant in the absence of predation, and densities of seedlings and saplings of canopy trees were greatly reduced, indicating the importance of predation to maintaining canopy community structure.

Various devices that exclude predators are the easiest methods for measuring predation rates. Cages or nylon sleeves over branches can be used to prevent access by birds and other large predators (Campbell and Torgersen 1983; Marquis and Whelan 1994; Mooney 2007). Sticky or teflon banding or aluminum baffles around branches or trees can restrict access by ants and other invertebrate predators, as well as nonflying predators (Campbell and Torgersen 1983; Dial and Roughgarden 2004; Mooney 2007). In a particularly ambitious study, Campbell et al. (1983) enclosed whole 7- to 12-meter-tall Douglas-fir trees with mesh to exclude birds.

Alternatively, gut contents of captured predators can be analyzed to estimate individual predation rates that can be scaled up to rates for the predator population (Reagan 1996; Stewart and Woolbright 1996; Waide 1996). This method works best when residence time of prey remains in predator guts can be estimated—that is, the time period represented by gut contents.

MEASUREMENT OF CHANGES THROUGH TIME

Canopy structure and conditions change through time as a result of changes in environmental conditions, disturbances, and tree growth. Diel variation reflects circadian patterns of plant and animal activity. Among the most important diel changes is the relative rates of photosynthesis and respiration that drive much of canopy structure and function. Many animals are active only during the day or night. Sampling techniques must address these diel patterns (e.g., the use of light traps or triggered flash photography for documentation of nocturnal activity).

Seasonal variation reflects phenology of growth and senescence relative to cold or dry periods. Many temperate and tropical forests are cold- or drought-deciduous (see Chapter 2). Such forests are foliated only during periods with adequate light, temperature, and moisture conditions. Many herbivores are attuned to seasonal availability of plant resources (e.g., newly flushed leaves or fruits and seeds). The scheduling of sampling activities should address both the timing of peak occurrence or activity and changes in demography as a result of geographic variation in availability of fruit and seed resources. Parmesan and Yohe (2003) conducted a global meta-analysis for 1,700 species and found a diagnostic 6.1 kilometer/decade range shift toward the poles (or a 6 meter/decade increase in elevation) and a 2.3 day/decade advance in spring events for 279 species (including plants, insects and vertebrates), indicating that climate change already is affecting ecosystems.

Documentation of changes in canopy conditions and diversity over time requires commitment to long-term sampling. Climate cycles, global climate change, frequency of storm, drought and fire disturbances, and canopy recovery following disturbance all occur over periods of many years to centuries. Platforms such as the network of International Long Term Ecological Research (ILTER) sites, canopy crane sites, and other secure, long-term research sites offer opportunities for long-term installation of canopy plots and sampling equipment unthreatened by changes in land use (see Chapter 3). Sampling of multiple canopy locations over time increases the likelihood of representing pre- and postdisturbance canopy conditions, as well as long-term responses to environmental change. However, long-term research must also be protected against changes in personnel and methods that would make portions of the dataset noncomparable for evaluation of trends. It is necessary to ensure that sampling methods are consistent so that data from later samples provide the same information as earlier samples.

EXPERIMENTAL METHODS

Much canopy research has described the diversity of canopy habitats and organisms. While such data have demonstrated the importance of forest canopies as reservoirs of a bulk of global biodiversity (Basset et al. 2007; Erwin 1982; Krömer et al. 2007; Obregon et al. 2011; Simmons and Voss 1998; Waide 1996), assessment of the critical habitat needs of various canopy species and the influence of various factors on their abundance and distribution requires appropriate experimental (manipulative) approaches that support evaluation and recommendations for conserving biodiversity (e.g., Dial and Roughgarden 1995; Hurlbert 1984; Lindo and Winchester 2007; Lowman 2009a; Mooney 2007; Shiels et al. 2010; Souza and Martins 2005; Whelan 2001). Furthermore, evaluating the effects of epiphytes, herbivores, and detritivores on canopy structure and habitat conditions also requires experimental approaches to support conclusions. However, whereas experimental manipulation has been used commonly to answer critical questions in rocky intertidal (Murray et al. 2006), grassland (e.g., McNaughton 1985; McNaughton 1986; McNaughton 1993) and desert ecosystems (Schowalter et al. 1999), forest canopies present special challenges associated with their three-dimensional complexity.

MANIPULATION OF CANOPY ENVIRONMENT

Experimental manipulation of the canopy environment is difficult for a variety of logistical reasons, discussed in Chapter 3. However, some experiments have manipulated branch density or orientation to evaluate effects on associated organisms. For example, Whelan (2001) experimentally manipulated leaf dispersion and attached artificial branches at different vertical spacing to demonstrate the effects of canopy architecture on the behavior of foliage-gleaning birds. De Souza and Martins (2005) attached artificial branches with high or low foliage density to branches of three tree species with similar foliage densities to isolate effects of foliage density on spider abundance. More spiders colonized artificial branches with high foliage density than branches with low foliage density. Roubik (1993) experimentally manipulated the height of floral resources from different canopy strata in tropical forests in Panama and demonstrated that the apparent fidelity of pollinator species to particular canopy strata reflected pollinator preferences for particular floral resources. Most pollinator species were attracted to their preferred floral resources regardless of their location in the canopy. Lindo et al. (2008) placed artificial soil habitats of different sizes at various heights within the canopy to assess the effect of soil volume, soil moisture, and canopy height on the development of oribatid assemblages (Fig. 5.2).

MANIPULATION OF CANOPY ORGANISMS

Experimental manipulation of canopy organisms or communities for evaluation of ecological roles is perhaps more difficult than manipulation of canopy structure (see Chapter 3). Manipulation of epiphyte or animal abundances is labor intensive and requires either manual removal and exclusion techniques or access to sufficient numbers (especially of insects), from either the field or colonies maintained in the laboratory, to increase abundance. Toledo-Aceves and Wolf (2008) planted seeds of an endangered canopy epiphyte, *Tillandsia eizii*, at different positions in tree canopies (e.g., inner canopy near the bole and middle canopy 2 m out from the bole) to assess differences in germination, survival, and growth in these different canopy positions. Addition of insects can be accomplished by transferring individuals or egg masses to experimental trees (Schowalter et al. 1991). The use of toxic or repellent chemicals to reduce herbivory can confound measurement of effects on nutrient fluxes, given the frequent incorporation of nitrogen or other important elements in these chemicals

FIGURE 5.2.
Artificial soil habitat in which soil volume and soil moisture can be manipulated to evaluate effects on arboreal flora and fauna.
Photo courtesy of Neville Winchester.

(Seastedt et al. 1983). Conversely, the simulation of herbivory by means of manual clipping or removing portions of vegetation may not represent all effects of herbivory, such as herbivore selection of particular foliage categories, effects of insect feces on plant production and carbon exchange, and so on. Consequently, experimental manipulation of canopy organisms generally has been limited to small trees (Frost and Hunter 2007; Frost and Hunter 2008; Kimmins 1972; Schowalter et al. 1991; Seastedt et al. 1983) or understory plants (Fonte and Schowalter 2005), on which numbers or density can be conveniently increased or decreased.

Experiments on small trees raise questions about the degree to which data from such studies represent processes in mature canopies. Several methods for reducing densities of target taxa have been used to evaluate effects on canopy structure or processes. Sexton and Schowalter (1991) compared the utility of sticky barriers and metal baffles around the main bole to exclude a flightless, ground-based herbivore from accessing tree crowns (Fig. 5.3). Dial and Roughgarden (1995) were able to remove *Anolis* lizards from isolated mature trees following Hurricane Hugo in Puerto Rico to test the effects of predation on canopy arthropods. Mooney (2007) used mesh fabric exclosures and sticky barriers to exclude birds and ants, respectively, from branches of mature ponderosa pines, *Pinus ponderosa*, to evaluate the effects of these predators on the canopy arthropod assemblage. Canopy detritivore density and the effects of litter decomposition can be manipulated by different litterbag mesh sizes or selective repellents that restrict access by particular taxa or size groups (Lindo and Winchester 2007).

However, experiments often incur confounding effects of variation in branch or overstory conditions, especially in the tropics where the canopy is a complex three-dimensional combination of various plant species. For example, the treatment of plots in the subcanopy or the forest floor to evaluate the effects of canopy herbivore-derived inputs on nutrient fluxes may be confounded by variation in throughfall and litterfall chemistry, resulting from the differential representation of tree species in the vertical column over each replicate plant or plot (Schowalter et al. 2012; Wullaert et al. 2009). This requires the strategic placement of experimental plots to minimize variation, large numbers of replicates to increase statistical power, and/or increased levels of experimental manipulation to ensure detection of effects (discussed later).

FIGURE 5.3.
Aluminum baffle used to restrict access to Douglas-fir trees by a flightless herbivorous weevil, *Lepesoma lecontei*.

Several statistical issues arise in selecting sampling methods. These include random versus systematic or stratified sampling, autocorrelation of samples, and the number of variables included in the analyses.

Many statistical analyses assume random sampling, which is largely impossible in forest canopies, given the limitations of canopy access and the nonrandom arrangement of plant modules. Climbing limits access to many potential sampling points at the periphery of the canopy, whereas towers and cranes limit access to many potential sampling points within the canopy. Much of the canopy volume is occupied by airspace between branches or foliage. Therefore, efforts must be taken to minimize bias, since accessibility often must be ensured at close proximity to a potential sample. Canopy cranes offer the best option for random sampling, since random x-, y-, and z-coordinates can be generated prior to sampling, and all accessible points from this pool are subsequently sampled without bias (e.g., Schowalter and Ganio 1998; Ernest et al. 2006). For other canopy access methods, a pool of random sample points can be created from a tree number, canopy aspect, and branch of a particular order or height, and the nearest accessible sample is then collected.

Independence of samples is a key assumption of statistical analyses, especially for testing hypotheses, for example, concerning effects of tree species, herbivory, or climate change (Hurlbert 1984). Sample independence requires that samples are not influenced by each other for treatment comparisons. For example, multiple samples within a tree crown are not independent, because any sample within a tree crown may be influenced by movement of materials from other samples upstream in the network of interconnected branches and foliage or by the effect of other sample units on habitat variables within the crown. For studies of vertical distribution, data can be pooled by height within a tree and tree treated as a block. However, for comparison of treatment effects on whole trees, all samples within each tree should be pooled, to provide one independent sample for comparison of variables among treated trees. Similarly, samples within plots are not independent with respect to comparisons among plots, since trees (and their crowns) within plots are subject to similar levels of environmental variables that follow geographic gradients. Therefore, samples within each plot also should be pooled for comparison of variables among plots.

Research in forests often has been undermined by "pseudoreplication," in which nonindependent samples have been treated as replicates for testing treatment effects (Hurlbert 1984). Examples include the use of a single pair of watersheds (one treated) for evaluation of the effect of canopy opening (Schowalter et al. 1981), a single pair of defoliated and nondefoliated sites for evaluation of the effect of defoliation on forest conditions, or a single pair of insecticide-treated or nontreated sites for evaluation of the effect of insect control. Multiple samples within each of the two sites were treated as replicates for statistical testing of the treatment effect, although any two plots may differ for reasons other than the obvious treatment. The consequence of reliance on such non-independent sampling may mask underlying factors that may determine the outcome. However, this situation may be remedied by combining multiple pair-plot studies into an integrated meta-analysis, with pooled data from each of the plots representing independent replication, or by sampling enough variables prior to treatment to ensure that experimental units are essentially identical prior to treatment.

A second way in which samples may not be independent is correlation through time. Samples collected at successive times are not independent to the extent that later samples are influenced by earlier sampling from the same location. For example, population size at time t + 1 is a function of population size at time t; destructive sampling at time t (e.g., by removing foliage) may alter conditions for samples collected at time t + 1. Because of access limitations, and especially if marked leaves or crowns are being monitored, repeated sampling of the same foliage, branch, or canopy location is common in canopy studies. Successive samples can be autocorrelated because of confounding effects of tree condition, spatial gradients, and so on, or by effects of previous sampling activity, and therefore not independent as required for statistical analyses. Data should be tested for significance of autocorrelation and adjusted as necessary to avoid unintentional confounding of results.

Another statistical issue involves analysis of nonrandom data sets, such as herbivory, for which data may be clumped at low and high leaf area missing (e.g., Lowman and Heatwole 1992). In many (but not all) cases, data can be normalized adequately through log transformation or other processes before analyses.

A frequent question is how best to represent the diversity of canopy flora and fauna. Relatively few studies have involved the variety of collection methods necessary to represent the entire diversity of canopy organisms. Studies of canopy invertebrates typically involve hundreds to thousands of species (Basset et al. 2007; Erwin 1982; Schowalter 1995; Schowalter and Ganio 2003; Lindo and Winchester 2007; Lindo et al. 2008). For example, Basset et al. (2007) used a variety of canopy access techniques and fourteen different sampling methods to sample arthropods in forest floor, understory, and mid- and upper-canopy habitats in a tropical rain forest in Panama (Investigating the Biodiversity of Soil and Canopy Arthropods [IBISCA] project). They collected approximately 500,000 arthropods belonging to about 5,500 species from 9,400 samples and 315 plant species. By contrast, studies of epiphytes and vertebrates typically deal with only a few dozen species. As a consequence, studies of flora and vertebrates can present species-level data relatively easily, but studies of invertebrates typically have either (1) focused on a few species of interest, such as a target herbivore and associated predators (Mooney 2007) or all Coleoptera (Grimbacher and Stork 2007), or (2) condensed taxa in various ways, particularly at higher ranks (e.g., family or order) or as functional groups (e.g., folivores, sap suckers, predators, and detritivores; Schowalter and Ganio 2003). Although limiting or condensing invertebrate diversity is more convenient, much information on diversity, especially as this affects responses to environmental changes or effects on canopy processes, is lost. Schowalter and Ganio (2003) and Schowalter et al. (1999) reported that species within a given family or functional group showed complementary responses to environmental changes—that is, some species increased in abundance whereas others decreased or showed nonlinear responses to environmental gradients. These species responses were masked at the family or functional group level, indicating that general patterns of community assembly are maintained across environmental changes through species replacement. Therefore, adequate representation of species and community responses to environmental change or treatment requires both (1) statistical analysis of individual species that are sufficiently abundant and (2) analysis of order or functional groupings for analysis of assemblage responses. Basset et al. (2008) compared diversity metrics for assessment of canopy arthropod responses to habitat disturbance and found that most metrics showed large differences between forest and nonforest habitats but were not equally discriminating for all taxa. Higher taxonomic groups were present in most habitats, but many insect species were more restricted to particular sites or habitats. Basset et al. (2008) concluded that metrics based on species identity had high sensitivity to disturbance, whereas measurements describing community structure were less discriminating. They also recommended using metrics based on abundance, species richness, species turnover estimated by multivariate analysis, and guild structure.

The interpretation of results is also affected by the level at which the data are analyzed. Analysis at the species level would provide the most detailed information, but few invertebrate taxa are sufficiently abundant for statistical analysis. Combining invertebrate species reduces the number of zeros among samples in a data matrix, improving the consistency of nonzero entries for statistical analyses.

Statistical analysis of a large number of variables increases the probability of type 1 error—that is, 5 percent of variables should show significant responses erroneously. For example, Progar et al. (1999) reported that two out of forty canopy arthropod taxa (i.e., 5 percent) showed significant responses to plot treatment designation prior to treatment implementation. However, the same error rate would be expected if species or other variables were analyzed individually, with no information on which variables showing significant responses were in error. The evaluation of multiple variables or taxa has the advantage of indicating general patterns (e.g., degree of consistency) of responses to treatments (e.g., Moran 2003). Furthermore, analyzing multiple taxa should reduce the probability of type 2 error (i.e., some taxa should show nonsignificant responses erroneously). If environmental changes or experimental treatments have general effects among taxa or functional groups, then at least some should show significant responses. Nevertheless, if fewer than 10 percent of variables show significant treatment effects, those effects should be interpreted with caution.

SUMMARY

Forest canopies present unique challenges to researchers in terms of access, location of sampling devices, and treatment replication. However, innovative methods have been developed to study canopy structure, internal conditions, community composition and function, and canopy effects on ecosystem processes, using a variety of access and sampling techniques.

Description of canopy structure is fundamental to understanding its effect on associated canopy organisms and on canopy interactions with atmospheric conditions. Forest canopies can be described as a network or radiating branches and foliage, with differences in branch orientation, density, and total and foliated length that largely determine habitat and resource conditions for canopy epiphytes and animals. Leaf area or mass determines carbon, water, and nutrient fluxes. Canopy epiphytes, invertebrates, and vertebrates can be surveyed to some extent from the ground, but many are most accurately sampled by observation or sampling devices placed in the canopy.

Forest canopies exchange energy, water, and nutrients with the atmosphere and geosphere in ways that affect regional and global climate and water yield and quality. Measurement of these processes and fluxes requires access methods for the placement of sophisticated tools in the canopy or remote-sensing technology capable of detecting fluxes from spectrometric data.

Experimental manipulation of canopy conditions, flora, and fauna has been particularly problematic. Most manipulations have been conducted in short trees, but some ambitious projects have manipulated canopy structure or canopy cover in relatively large plots of forest or installed pumps and tubing to increase atmospheric CO_2 concentrations.

In addition to logistic problems of experimental design and manipulation, canopy studies must address several statistical issues, including randomization of sampling, sample independence, and level of taxonomic discrimination in statistical analyses. The selection of statistically independent sample units is particularly problematic, given the difficulty of replicated access to and manipulation of forest canopies. The level of taxonomic discrimination is an issue for invertebrate studies, which can include hundreds of species—few of which are sufficiently abundant for statistical analysis. A practical solution is to analyze sufficiently abundant species individually, as well as in combination with other species, in order to assess species and community responses to environmental changes or treatments.

SUGGESTED READING

Arriaga-Weiss, S. L., S. Calmé, and C. Kampichler. 2008. Bird communities in rainforest fragments: Guild responses to habitat variables in Tabasco, Mexico. *Biodiversity Conservation* 17: 173–90.

Dial, R., M. D. F. Ellwood, E. C. Turner, and W. A. Foster. 2006. Arthropod abundance, canopy structure, and microclimate in a Bornean lowland tropical rain forest. *Biotropica* 38: 643–52.

Díaz, I. A., K. E. Sieving, M. E. Peña-Foxon, J. Larraín, and J. Armesto. 2010. Epiphyte diversity and biomass loads of canopy emergent trees in Chilean temperate rain forests: A neglected functional component. *Forest Ecology and Management* 259: 1490–501.

Ellwood, M. D. F., D. T. Jones, and W. A. Foster. 2002. Canopy ferns in lowland dipterocarp forest support a prolific abundance of ants, termites, and other invertebrates. *Biotropica* 34: 575–83.

Lowman, M. D., R. L. Kitching, and G. Carruthers. 1996. Arthropod sampling in Australian subtropical rain forest: How accurate are some of the more common techniques? *Selbyana* 17: 36–42.

Malcolm, J. R. 2004. Ecology and conservation of canopy mammals. In *Forest canopies*, 2nd edition, ed. M. D. Lowman and H. B. Rinker, 297–331. San Diego, CA: Elsevier/Academic Press.

Oliveira-Santos, L. G., M. A. Tortato, and M. E. Graipel. 2008. Activity pattern of Atlantic forest small mammals as revealed by camera traps. *Journal of Tropical Ecology* 24: 563–67.

Ozanne, C. M. P. 2005. Techniques and methods for sampling canopy insects. In *Insect sampling in forest ecosystems*, ed. S. R. Leather, 146–65. Malden, MA: Blackwell/J. Wiley and Sons.

Schowalter, T. D., and L. M. Ganio. 1998. Vertical and seasonal variation in canopy arthropod communities in an old-growth conifer forest in southwestern Washington, USA. *Bulletin of Entomological Research* 88: 633–40.

6

CANOPY–ATMOSPHERE
INTERACTIONS

Forest canopies interact with the atmosphere in a number of ways that affect local and global climate and exchange of carbon and other materials. Forest canopies buffer terrestrial surfaces from extreme variation in temperature, precipitation, and wind, moderating conditions for associated organisms and ecological processes. The degree of buffering depends on the extent of canopy cover, depth, and complexity. Exchange of materials, especially water, occurs through foliage stomata, influencing local humidity and precipitation patterns. Deforestation has disrupted canopy–atmosphere interactions, exposing the land surface to greater variation in regional temperature, precipitation, and wind speed (Foley et al. 2003).

These interactions can be measured by means of data collected above, within, and below the canopy to measure the effects of the canopy as a source (producer) and sink (recipient) of materials (Carter and Knapp 2001; Foley et al. 2003; Fonte and Schowalter 2005; Frost and Hunter 2004; Frost and Hunter 2007; Frost and Hunter 2008; Reynolds and Hunter 2001; Schowalter et al. 1991; Seastedt et al. 1983). The adequate representation of canopy interaction with the atmosphere, as affected by climate, disturbance, or herbivory, requires multidisciplinary approaches, involving meteorologists, biogeochemists, and ecologists, and measurement of multiple factors. Modern remote-sensing equipment and eddy covariance techniques have greatly simplified the effort required to study canopy–atmosphere interaction and led to major advances in our understanding of these interactions. This chapter focuses on processes of canopy–atmosphere interaction, effects on climate, and methods for experimental manipulation.

CANOPY INTERACTION WITH THE ATMOSPHERE

Forest canopies represent an important biotic machine that stores and moves energy, water, and nutrients among geospheric, biospheric, and atmospheric pools. Photosynthesis and respiration significantly affect carbon exchange with the atmosphere (Foley et al. 2003). Reduced albedo and evapotranspiration by the canopy moderate local temperature (Foley et al. 2003). Interception and evapotranspiration of precipitation by forest canopies control the movement of water, cloud formation, and future precipitation (Juang et al. 2007). Finally, canopy disturbances and herbivory, decomposition, and predation in forest canopies affect foliage turnover and fluxes of energy, water, and nutrients through the forest food web (Schowalter et al. 1991; Seastedt et al. 1983).

Photosynthesis is the engine that stores energy and carbohydrates to drive all other ecosystem functions, as well as ecosystem services valued by humans, and is a primary function of forest canopies. Respiration reverses this process, as the energy of stored carbohydrates is released to perform the various metabolic functions of trees and the community of organisms in the canopy. Fluxes of these and other biogenic gases affect carbon storage and distribution in the canopy and influence atmospheric conditions (Lerdau and Throop 1999; Misson et al. 2007; Turner et al. 2004; Turner et al. 2005; Turner et al. 2007). Therefore, measurement of these functions is a fundamental activity of canopy ecologists, who have shown great creativity in developing tools for measuring photosynthesis and respiration.

Forest canopies provide a large surface area of branches and foliage for interception of precipitation and airflow. Canopy height, canopy cover, and vegetation type determine how much precipitation is intercepted before reaching the forest floor (Juang et al. 2007). In montane forests, taller, denser canopies can acquire water directly by interception of rising clouds, augmenting annual precipitation (Brauman et al. 2010). Deeper and denser canopies also intercept more precipitation than shorter and sparser canopies. Epiphytes increase water interception and storage (Pypker et al. 2006). Dry deposition of particulate materials and nutrients is often a substantial portion of total atmospheric inputs to forest canopies (Lovett and Lindberg 1993).

Interception of precipitation channels and stores water and dissolved nutrients in canopy reservoirs (such as tree holes and phytotelmata) and reduces the volume and impact of water reaching the forest floor, thereby reducing erosion and facilitating infiltration and storage in litter and soil. Interception of airflow results in reduced wind speed and absorption and adsorption of airborne particulates and nutrients, augmenting resource inputs in precipitation and storage in soil. Evapotranspiration helps drive water translocation in tall trees and also contributes to canopy cooling and to precipitation of locally generated condensed moisture (Foley et al. 2003; Juang et al. 2007). These canopy functions therefore play key roles in forest hydrology and nutrient fluxes.

Forest canopies intercept airflow, reducing wind speed, creating turbulence (as described previously), and acquiring particles and aerosols from the air. Reduced airflow affects canopy gradients in temperature, relative humidity, and consequently evapotranspiration rate. Dry deposition of particles (adsorption) and absorption of aerosols provides sediment and material that enhance canopy function (soil development and nutrient input) or stress plants and interfere with canopy function (pollutants). For example, Solberg et al. (2009) reported that European forests, especially pine and spruce forests, have shown greater-than-predicted growth rates over the past 15 years, explained largely by a fertilization effect of atmospheric nitrogen deposition.

MODIFICATION OF ENVIRONMENTAL CONDITIONS

Forest canopies provide considerable capacity to modify environmental conditions, maintaining relatively stable conditions within forest ecosystems. Forest canopies emit a variety of volatile compounds that affect atmospheric chemistry. In addition, forest canopies modify local and regional climates and thereby affect global climate. Furthermore, canopy conditions buffer forests against many disturbances.

Forests generate a number of volatile organic compounds (VOCs) that influence atmospheric chemistry, especially oxidative potential (Guenther et al. 1996; Heald et al. 2009; Lelieveld et al. 2008). The most abundant VOCs are isoprene and monoterpenes (Lerdau et al. 1997). Tropical forests are the largest source of isoprene (Lelieveld et al. 2008; Lerdau and Throop 1999). Isoprene biosynthesis and emission rates are related to light intensity and temperature and foliar nitrogen availability, whereas monoterpene biosynthesis and emission rates appear to be controlled primarily by temperature (Harley et al. 1996, 1997; Heald et al. 2009; Lerdau et al. 1997; Pressley et al. 2006). Methanol and acetone also are emitted by many plant species. Canopy plants and animals emit additional volatile compounds (e.g., plant signaling and defensive compounds and animal pheromones) that have relatively little effect on atmospheric conditions but are critical to regulating species interactions (Cardé and Baker 1984; Dolch and Tscharntke 2000; Lerdau et al. 1997; Murlis et al. 1992).

A major effect of these compounds is their light-sensitive oxidation into hydroxyl radicals, ozone, and carbon monoxide (Lerdau et al. 1997; Lelieveld et al. 2008). Carbon monoxide, in particular, influences the oxidizing capacity of the atmosphere and is involved in photochemical reactions that increase atmospheric ozone concentration.

However, isoprene also functions to increase the longevity of methane in the atmosphere, thereby indirectly contributing to global warming (Lerdau and Throop 1999; Lerdau et al. 1997). Background isoprene emission by forests appears to be in balance with atmospheric oxidative capacity and may function to maintain atmospheric conditions conducive to forest production, but deforestation and conversion to agricultural or urban uses is likely to disrupt this balance (Lelieveld et al. 2008).

Forest canopies modify canopy conditions in several ways. They shade and protect the forest floor, abate airflow, and modify precipitation (Foley et al. 2003; Janssen et al. 2008; Juang et al. 2007; Madigosky 2004; Parker 1995), maintaining more moderate temperatures and relative humidities than occur in more open environments. Volatile organic compounds act as greenhouse gases and warm the atmosphere (Lelieveld et al. 2008). These effects buffer forests against much variation in ambient climate conditions, as well as many disturbances (see next section).

The forest canopy absorbs solar energy and reflects light and heat, lowering albedo and reducing surface temperatures (Foley et al. 2003; Gash and Shuttleworth 1991; Lewis 1998). Albedo is inversely related to canopy height and "roughness" (the degree of unevenness in canopy surface), declining from 0.25 for canopies less than 1 meter in height to 0.10 for canopies more than 30 meters in height, and reaches lowest values in tropical forests with very uneven canopy surfaces (Monteith 1973). Canopy roughness also generates turbulence in airflow (Cassiani et al. 2008; Raupach et al. 1996; Su et al. 2008), thereby contributing to surface cooling by wind (sensible heat loss), evapotranspiration (latent heat loss), and the rise of moist air to altitudes at which condensation and precipitation occur (Foley et al. 2003; Meher-Homji 1991). At night, the canopy absorbs reradiated infrared energy from the ground, maintaining warmer nocturnal temperatures than deforested sites.

Evapotranspiration contributes to canopy cooling and to convection-generated condensation above the canopy, thereby increasing local precipitation (Foley et al. 2003; Juang et al. 2007; Meher-Homji 1991). Evapotranspiration increases relative humidity above the canopy and, coupled with strong advective moisture flux, especially in the tropics, promotes local cloud formation (Trenberth 1999). Furthermore, volatile chemicals emitted from canopy foliage can serve as precipitation nuclei (Facchini et al. 1999). Canopy removal over large areas (i.e., deforestation) has been associated with declining local and regional precipitation (Janssen et al. 2008; Meher-Homji 1991) due to positive feedback between reduced canopy cover, increased albedo, and regional drying.

Forests are subject to natural disturbances, such as windstorms, fire, flooding, and drought, as well as anthropogenic disturbances, such as the harvest and alteration of forest structure and composition for commercial purposes (see Chapter 2). Canopy abatement of wind effects depends on tree structure, canopy density, and wind speed. Individual tree structure (e.g., height, taper, rooting depth or other buttressing, wood density, and branching pattern) affects sway frequency (oscillations per minute) and damping ratio (ability to return to resting position) when exposed to wind (Moore and Maguire

2005). Trees surrounded by other trees are buffered from wind effects, but more isolated trees experience the full effect of wind. Trees with sway frequencies that match wind gust frequency are more likely to experience damaging increase in sway. Branch and foliage density affect wind resistance. Rapid foliage loss during high winds reduces wind resistance. Neighboring trees in dense forests buffer wind force, whereas isolated trees experience the full force of the wind.

Cyclonic storms (hurricanes, typhoons, other tropical storms, and tornadoes) represent the greatest challenges to forests. Walker et al. (1991) described the effects of Hurricane Hugo, with wind speeds greater than 166 kilometers/hour, on tropical rain forests in Puerto Rico in 1989. Although the forest was nearly entirely defoliated, and many overstory trees snapped or were uprooted, tree mortality was relatively low and concentrated on exposed ridges and windward slopes (Brokaw and Grear 1991; Walker 1991). Most subcanopy and understory trees were relatively undamaged. Walker (1991) reported rapid refoliation and sprouting from broken and fallen boles; *Cecropia schrebe-riana* and other pioneer species rapidly filled canopy gaps. Canopy function was largely restored within five years, but at reduced average height.

Fire severity depends primarily on the amount and moisture content of litter fuels and canopy depth and structure. Fire occurs most frequently in relatively arid, open-canopied forests dominated by fire-tolerant plant species (e.g., trees that have insulating bark and shed lower branches as they grow; Agee 1993). Under these conditions, fires have relatively low heat and flame height and are largely restricted to burning fine litter and scattered coarse woody debris on the ground. Forest-replacing fires are more likely to reach tree crowns in dense, multicanopied forests with abundant dry litter. Under these conditions, fires easily climb into the canopy and can generate sufficient heat to create cyclonic firestorms that desiccate forests in advance of the fire, increasing severity and scale of disturbance. However, dense canopies can protect soil/litter moisture to a large extent (as mentioned previously) and thereby prevent or minimize fire until litter dries out.

Flooding is most frequent in riparian or low-lying forests, which typically experience seasonal flooding. Effects depend on duration of flooding, tree height, root stability or buttressing, and the ability to tolerate soil saturation and hypoxia (e.g., Sojka 1999). Flooding reduces soil strength, increasing the likelihood of uprooting of trees, and greatly slows oxygen diffusion in the root zone, stopping potassium uptake and causing stomatal closure in the canopy (Sojka 1999). However, evapotranspiration from the canopy moves considerable water up and out of the soil. Seasonal flooding in the Amazon basin is an important mechanism for distributing seeds and for initiating succession.

Drought occurs with some frequency in virtually all forests. Even tropical rain forests are subject to periodic drought, especially resulting from changes in global circulation patterns (e.g., El Niño; Windsor 1990). For example, during a prolonged drought in Puerto Rico during 1993–95, the rain forest received only 63 percent of the annual mean precipitation (Heartsill-Scalley et al. 2007). Although forest canopies can protect soil/litter moisture to a point, water limitation causes disruption of capillarity, cell cavitation, and death of canopy parts deprived of water (see Chapters 2 and 3). However, drought effects are more severe for overstory trees because of the greater exposure to desiccation and greater difficulty in transporting water for tall trees. Nepstad et al. (2008) reported that mortality increased 4.5 times among large trees and 2 times among medium-sized trees after 3 years of experimental reduction, by 60 percent, of incoming throughfall. Although such drought effects change forest structure and composition for long periods of time, overall canopy function may be largely maintained by the surviving subcanopy. Furthermore, drought frequently triggers outbreaks of herbivorous insects (Mattson and Haack 1987; Van Bael et al. 2004). Although outbreaks often are viewed as disturbances, or at least as exacerbating disturbances, their effects are more complex and include reduced water demand and increased survival (at least of seedlings) during drought (Kolb et al. 1999; Sthultz et al. 2009) and selective reduction of stressed or abundant plant species, that may tailor the postoutbreak community to prevailing conditions (Schowalter 2011). In this way, the canopy community may function to mitigate drought effects.

Anthropogenic disturbances have variable effects, depending on their similarity to natural disturbances. However, forest harvest typically creates patterns in canopy cover that differ substantially from patterns in natural forests. In particular, harvest creates distinct edges that contrast with the broader ecotones among forest types or successional stages in natural forests. These edges, in turn, result in warmer, drier conditions that characterize clear-cuts influencing environmental conditions several hundred meters into the neighboring forest (Chen et al. 1995). In some cases, manipulation of disturbance regime results in changes in forest conditions that can increase the severity of disturbance. For example, the semiarid, open-canopied pine forests that historically dominated western North America were characterized by frequent, low-intensity fires. Increased density and canopy closure by fire-intolerant firs in these forests, as a result of fire suppression during the twentieth century, resulted in extensive mortality to water-limited firs and, eventually, catastrophic fire (Schowalter 2008).

EFFECTS OF CANOPY ASSOCIATES

Associated organisms can affect these canopy–atmosphere interactions in a number of ways. Some species enhance, while others interfere with, these interactions.

Epiphytes can contribute to canopy moderation of forest microclimates. Stuntz et al. (2002) found that epiphytes significantly reduced midday temperatures and evaporative drying in the surrounding canopy, even compared to branches within the same tree that were devoid of epiphytes. Classen et al. (2005) demonstrated that herbivory by both sap-sucking and folivorous insects reduced foliage density enough to reduce crown interception of precipitation and affect temperature and relative humidity around treated trees, enough to affect ecosystem processes.

MEASUREMENT OF CANOPY–ATMOSPHERE INTERACTIONS

VOLATILE ORGANIC COMPOUNDS

Fluxes of volatile organic compounds can be measured by means of a gas-exchange system with a temperature- and light-controlled cuvette that routes sampled air directly to a gas chromatograph with a photoionization detector (Lerdau and Throop 1999). Pressley et al. (2006) described the measurement of isoprene emission from towers in a hardwood forest. Air from the 31-meter level was delivered at a high rate of turbulent flow (Re = 8,043) to an open-path infrared gas analyzer at the base of the tower via 1.27-centimeter inside diameter Teflon tubing. The lag time between the inlet at 31 meters and the sensor at the base of the tower averaged 10 seconds. All measurements were collected at 10 hertz and processed offline to generate 30-minute average fluxes. Density corrections were applied according to Webb et al. (1980) and high frequency attenuations in the isoprene flux were accounted for by applying a low-pass filter to the sensible heat flux to determine a correction based on the ratio of the unfiltered to filtered sensible heat flux (Massman 2000).

CLIMATE MODIFICATION

Measuring canopy effects on forest microclimate requires measurement of temperature, moisture, and wind speeds above, within, and below the canopy and among forested and nonforested sites using standard meteorological instruments placed at designated heights (Madigosky 2004; Parker 1995; Sillett and Van Pelt 2007). Light sensors, temperature/humidity probes, rain gauges, and three-dimensional anemometers, for example, can be located at height intervals on towers or tree boles (Fig. 6.1). However, measurement of ambient temperature, radiative heat flux, horizontal and vertical airflow, eddy covariance, relative humidity, and precipitation requires the placement of instruments at fixed heights above the canopy, as well as within and below the canopy (Funk and Lerdau 2004). Equipping sensor arrays with solar power sources facilitates continuous monitoring (Fig. 6.1D; Sillett and Van Pelt 2007).

Similar arrays placed in replicated canopy gaps or among forest types or successional stages can be used for comparison to evaluate effects of forest canopy structure on climate variables over larger areas. As described in Chapter 4, appropriate independent replication of treatments among geographically intermixed plots is necessary for the most robust statistical analysis to support policy decisions.

MODIFICATION OF DISTURBANCE

Disturbance magnitude and severity can be measured using standard methodology. Wind speed is measured with anemometers at selected heights in and above the canopy. Degree of tree sway and potential dislodging of root systems can be measured with

FIGURE 6.1.

(A) Instrumentation at height increments on tower for measurement of eddy covariance,

(B) instruments mounted above canopy for measurement of canopy–atmosphere interaction,

(C) detail of anemometer for measurement of wind speed and eddy covariance, and

(D) solar panel providing power for remote instrumentation.

Photo A courtesy of Beverly Law.

strain transducers (Moore and Maguire 2005). Extent of defoliation, stem breakage, and uprooting provide measures of storm severity. Fire intensity and severity can be determined by thermisters in the soil and litter or by measuring scorch height on boles and branches and the extent of charred boles and stumps. Drought severity can be measured as rate of cell cavitation, using acoustic equipment attached to boles, as well as by measurement of extent of canopy death (Mattson and Haack 1987).

PHOTOSYNTHESIS AND RESPIRATION

The most ambitious early study (Odum and Jordan 1970) involved enclosing a multi-tree patch of tropical rain forest in a plastic bag, with provision for necessary ventilation, and measuring whole-canopy photosynthesis, respiration, and net primary production (Fig. 6.2). To our knowledge, this feat has not been repeated. Most studies of photosynthesis have involved enclosure of individual leaves or foliage-bearing branches in a gas-exchange analyzer (Funk and Lerdau 2004; Irvine et al. 2005; Lerdau and Throop 1999) and scaling the results to the whole-tree and whole-canopy levels (Funk

FIGURE 6.2.
Giant plastic cylinder used to measure photosynthesis and evaporation by a tropical rain forest in Puerto Rico. Top left, diagram of the cylinder showing position of instruments; bottom left, inside lateral view of vegetation and partially raised plastic enclosure during construction; right, aerial view of the cylinder with the plastic raised to the 17-meter level.
From Odum and Jordan (1970).

and Lerdau 2004; Law et al. 2006; see Chapter 4). However, enclosures alter the conditions affecting photosynthesis and respiration, and such measurements represent only short-term conditions (Baldocchi et al. 1988).

Baldocchi et al. (1988) helped pioneer micrometeorological (eddy covariance) methods for studying canopy–atmosphere interactions (Box 6.1). Their concept was that the rate of change in chemical composition of the atmosphere at a point of measurement is balanced by mean vertical and horizontal advection, by mean vertical and horizontal divergence or convergence of turbulent flux, by molecular diffusion, and by any source or sink. Advection is defined as the mean transport of an atmospheric property by mean motion of the atmosphere. Convergent flux results in accumulation of material in a controlled volume (because flux entering is greater than flux exiting); divergent flux results in loss of material from the volume. Molecular diffusion in the atmosphere generally is negligible compared to turbulent transfer. The rates of photosynthesis and respiration are examples of a sink and a source of CO_2, respectively.

BOX 6.1: MICROMETEOROLOGICAL MEASUREMENTS OF ECOSYSTEM PROCESSES

Beverly Law

Micrometeorological measurements are made above canopies of different biomes to quantify and understand the exchanges of carbon dioxide, water vapor, and energy across a range of disturbance histories and climatic conditions. The micrometeorological eddy covariance technique provides a direct measure of net carbon uptake integrated over ~1 km². These measurements are computed half-hourly and are commonly summarized annually. This allows for the determination of seasonal and interannual variation in net carbon uptake, associated with climate variability and disturbances.

Clusters of flux sites in the same forest type at different developmental stages have been used to determine the effects of disturbance on forest processes by trading space for time (chronosequences). The climate and soil conditions are similar, so that differences are associated with disturbances (i.e., physiological responses) that differ with stand age, legacy carbon (coarse and fine woody debris), and successional understory vegetation responses. A synthesis of disturbance clusters in forests across North America examined the effects of stand-replacing disturbance from fire, clearcut harvest, and insects on carbon uptake. The study found that net ecosystem production (NEP) showed a carbon loss from all ecosystems following a stand-replacing disturbance, becoming a carbon sink in twenty years (Amiro et al. 2010), somewhat consistent with a global study over temperate and boreal forests that showed a transition from source to sink at an average of fifteen years (Luyssaert et al. 2008). In addition, NEP of boreal forests was relatively time-invariant after twenty years, whereas western forests tended to increase in carbon gain over a much longer time frame. This was driven mostly by gross photosynthetic production (GPP) because total ecosystem respiration (ER) and heterotrophic respiration didn't change much with age. One of the chronosequences in Oregon semiarid ponderosa pine showed that an old-growth forest was still an annual carbon sink (Law et al. 2002), and the early results at this site as well as other old forests globally indicated that most of the old forests were not carbon neutral (neutrality being where photosynthesis equals ecosystem respiration), contrary to the long-held theory from an ecological plantation study (Luyssaert et al. 2008). Many process models assume carbon neutrality in old forests, which would lead to underestimates of carbon uptake for about 15 percent of forests globally. Research at the ponderosa pine site in Oregon also showed the strong linkage between transpiration, gross photosynthesis, and root respiration during the growing season and a longer-term correlation between basal area increment and the base rate of soil respiration, providing compelling evidence for better linkages between canopy and soil processes in ecosystem models (Irvine et al. 2008).

The data from flux sites are important for calibrating ecosystem models. A report of the National Academies of Science in the United States was commissioned to provide recommendations for improving estimates of greenhouse gas emissions (NRC 2010). The report recommended integration of remote-sensing data and flux-site observations to reduce uncertainty in regional analyses of net annual carbon uptake by terrestrial ecosystems. As part of the North American Carbon Program, a West Coast US project demonstrated the integration of remote-sensing data on disturbance, forest age, and land cover; AmeriFlux data; and inventories for mapping net carbon uptake for every square kilometer (Law et al. 2006). The study used Landsat time-series analyses and ancillary data to identify disturbance type and year over the Landsat record (thirty years; Law et al. 2004). The AmeriFlux chronosequence data in different age ponderosa pine stands indicated that the model was underestimating carbon uptake in the old forests, so the data were used to improve the model estimates. Inventory data were used to change carbon allocation with stand age in the model. Thus the flux data prove useful in providing the net of multiple ecosystem processes that are simulated in the model, and combined with ancillary data, like above-ground production and transpiration, the flux site data are valuable for diagnosing weaknesses in model assumptions.

(continued)

(continued)

FLUX MEASUREMENTS AND INSTRUMENTATION

A critical aspect of using the eddy covariance (EC) method is appropriate placement of the tower. The method was developed for measuring mass and energy exchange over a horizontal surface with uniform vegetation canopies and flat terrain. The spatial scale of influence of ecosystem processes on the observations at a tower extends through the footprint around the tower (between 100 m for short canopies and 1,000 m for forests), so uniform vegetation (e.g., same forest stand height and density) over flat terrain should be sought over at least a kilometer for forests. A goal is to minimize horizontal flux divergence and horizontal and vertical advection (assumed to be negligible). For example, cold-air drainage or flux divergence at night can cause nighttime respiration to be greatly underestimated and thus net ecosystem exchange of CO_2 (the net of photosynthesis and respiration). Footprint analysis is used to determine the source area under a range of atmospheric stability conditions to provide guidance for placement (Göckede et al. 2008).

The EC instruments used are a fast-response (10–20 Hz) infrared gas analyzer (closed or open path IRGA) that measures CO_2 and H_2O, and a three-dimensional sonic anemometer-thermometer. The open-path analyzers do not operate when the window is wet or icy, so there is loss of data under these conditions, and the closed-path analyzers have inlet tubes that need to be cleaned to minimize attenuation of high-frequency components. The placement of the instruments is also critical. The sonic anemometer should be level, to minimize uncertainties in wind direction, and placed on a stable boom facing the mean wind direction to minimize flow distortions. The closed path inlet tube of the IRGA is placed close to the anemometer, so both instruments are essentially sampling the same eddy. The open-path IRGA should be placed next to the sonic anemometer in a way that does not cause flow distortion, and oriented at fifteen to thirty degrees to the horizontal plane, which helps drain moisture from the lower window. Details of instrument and tower placement and instrument manufacturers can be found on the AmeriFlux website (http://public.ornl.gov/ameriflux). Additional instruments are recommended for CO_2 and temperature profiles, particularly in tall forest canopies, as the CO_2 profile data are used to produce estimates of CO_2 storage in the canopy airspace at night and the temperature profiles provide diagnostic information (e.g., on potential flux divergence). Details of flux analysis and diagnostics can be found in a micrometeorology book (Massman et al. 2004).

The mean covariances between wind velocity components (streamwise, lateral, and vertical) and streamwise direction represent turbulent fluxes. The mean vertical turbulent flux of material (F) over a horizontally homogeneous surface under steady-state conditions is

$$F = -\rho_a w'\chi',$$

where ρ_a is the density of dry air, w' is the fluctuation in vertical wind flux, and χ' is change in mean direction (Baldocchi et al. 1988). Flux is directed downward when $F < 0$ and upward when $F > 0$. The simplest flux representation would be over relatively homogeneous surfaces that generate little turbulence. However, complex terrain and, especially, tall rough canopy surfaces generate considerable wake turbulence due to their interaction with atmospheric turbulence, requiring that measurements be made above the zone affected by canopy-generated turbulent eddies. Furthermore, measurement of gas fluxes can be influenced by chemical reactions occurring between the point of measurement and the surface. Therefore, fluxes of primary constituents and reacting compounds, as well as the compound of interest, should be measured at two heights to account for flux divergence (Baldocchi et al. 1988).

Eddy covariance relies on accurate measurement of vertical movement of air and materials, using a three-dimensional sonic anemometer. Accurate adjustment of sensor orientation is critical. Baldocchi et al. (1988) provided remedies for common sources of measurement error, such as sensor response time, distance between sensors, and sensor height and orientation (improper sensor angle), and listed typical sensors and micrometeorological techniques used to measure turbulent fluxes of various atmospheric chemicals, such as infrared gas analyzers for water and CO_2 fluxes (Table 6.1). Eddy-covariance measurements from various heights on towers have been particularly useful in providing estimates of carbon exchange rates in different canopy layers or successional stages (Baldocchi et al. 1988; Chen et al. 2002; Leuning et al. 2008; Misson et al. 2007), as well as fluxes in water and nutrients. Baldocchi et al. (2000) applied vertical advection theory to tall forests in complex terrain and concluded that inclusion of the vertical advection correction improved calculation of carbon exchange during the growing season but not during winter. Xiao et al. (2008) demonstrated that eddy flux measurements provided for accurate scaling from local to continental estimation of net ecosystem carbon exchange. Current methods for eddy covariance measurements are described by Massman et al. (2004; Table 6.1).

More recently, remotely sensed spectrophotometric data have been used to measure net carbon storage and flux of forest canopies (Irvine et al. 2002; Irvine et al. 2005; Law 2005; Law et al. 2006; Turner et al. 2004; Turner et al. 2005; Turner et al. 2007). Isotopic ratios change as gases are enriched or depleted during passage among biotic and abiotic pools, providing a measure of carbon flux. Bowling et al. (2003) measured $\delta^{18}O$ of precipitation, soil and leaf water, and respired CO_2 to evaluate the degree to which the isotopic enrichment of leaf water (through evapotranspiration) influences ecosystem respiration.

TABLE 6.1. Examples of Chemical Sensors and Micrometeorological
Techniques Used to Measure Turbulent Fluxes

Chemical	Sensor	Technique
H_2O	Lyman-alpha hygrometry	Eddy correlation
	Infrared absorption	Eddy correlation or flux gradient
CO_2	Infrared absorption	Eddy correlation or flux gradient
	Chemical absorption	Flux gradient or mass balance
SO_2	Flame photometry	Eddy correlation
	UV absorption	Eddy correlation
	Bubblers	Flux gradient
NO	O_3 luminescence	Eddy correlation
NO_2	O_3 luminescence	Eddy correlation
	Luminal	Eddy correlation
O3	NO luminescence	Eddy correlation
	Bubbler	Flux gradient
HNO_3	Nylon filters	Flux gradient
NH_3	Absorption traps	Flux gradient or mass balance
CH_4	Grab-bag air sample	Flux gradient
	Bubblers	Flux gradient

SOURCE: From Baldocchi et al. (1988), with permission from the Ecological Society of America.

Another ambitious project is the Amazonian Tall Tower Observatory (ATTO) being constructed near the Amazon River in Brazil (Tollefson 2010). The 320-meter-tall tower, rising far above the 35-meter-tall canopy, will provide large scale measurement of atmospheric conditions over much of the eastern Amazon, representing about 50 percent of the Amazon basin.

PRECIPITATION INTERCEPTION AND EVAPOTRANSPIRATION

Interception of precipitation and dissolved nutrients is most easily measured by comparing the volume and chemical composition of precipitation above and below the canopy, with differences representing the effect of interception by the canopy (McDowell 1998; Oishi et al. 2008; Seastedt et al. 1983; Schowalter et al. 1991; Solberg et al. 2009).

Collecting bottles, often with a funnel to increase collection area, can be suspended at various heights through the canopy and on the forest floor to measure precipitation volume for a given horizontal area. Because some branching patterns channel water toward either the bole or the crown perimeter, collectors should be placed either systematically at designated radial distances from the bole or randomly to account for such variation. Collectors can be attached to the bole to collect and measure precipitation channeled down the bole as stemflow (Schowalter et al. 1991).

Evapotranspiration maintains water supply to the canopy and influences relative humidity and water flux between the canopy and atmosphere. Evapotranspiration is a function primarily of latent heat flux (LE), derived from temperature, wind speed, stomatal conductance, and vapor pressure deficit (Irvine et al. 2002; Leuzinger and Körner 2007; Novick et al. 2009). The evapotranspiration rate can be calculated from canopy-level sap flux (Oishi et al. 2008; see Chapter 7) but is most often calculated by the measurement of latent heat flux via eddy covariance methods (Novick et al. 2009; Oishi et al. 2008; see above) and converted to evapotranspiration rate, using the temperature-dependent latent heat of vaporization for water vapor. Evapotranspiration ($kg\ m^{-2}\ sec^{-1}$) is equivalent to λ^{-1} multiplied by the latent heat flux to the atmosphere, where λ is the latent heat of vaporization ($J\ kg^{-1}$; Giambelluca et al. 2009). Evapotranspiration (E_c) also can be calculated as

$$E_c = LE - I_c - E_s,$$

where I_c is canopy interception and E_s is evaporation from the forest floor (Oishi et al. 2008).

Stomatal conductance is an important indicator of plant water status, as well as a regulator of evapotranspiration. Mean midday stomatal conductance (g_c, in $m\ sec^{-1}$) can be calculated as

$$g_c = E_T\ \frac{(\gamma\lambda)}{c_p\,\rho_a}\ \frac{(1)}{D}\ ,$$

where E_T is canopy evapotranspiration ($Kg\ m^2\ sec^{-1}$), γ is the psychrometric constant ($Pa\ K^{-1}$), λ is the latent heat of vaporization of water ($J\ kg^{-1}$), c_p is the specific heat of air ($J\ kg^{-1}\ K^{-1}$), ρ_a is the density of dry air ($kg\ m^{-3}$), and D is the saturated vapor pressure deficit of air (Pa; Irvine et al. 2002).

Although evapotranspiration is greatest during the daytime, nocturnal evapotranspiration is significant and should not be ignored (Novick et al. 2009). Variables used to calculate evapotranspiration generally have been measured by instruments mounted on towers extending above the canopy, but Zhang et al. (2009) demonstrated that evapotranspiration could be predicted accurately from satellite (MODIS) data on gross primary production and water-use efficiency.

Measurement of evapotranspiration using eddy-covariance involves a three-dimensional sonic anemometer (TSA) and open-path infrared gas analyzer (IRGA) positioned above the forest canopy (Oishi et al. 2008). Vertical temperature, relative humidity, and wind velocity are recorded and averaged using a data logger. Rain may block the path between transducers in the TSA or optical length in the open-path IRGA, requiring identification of such data gaps to ensure correct long-term sums and averages (Falge et al. 2001; Oishi et al. 2008). Oishi et al. (2008) provided a list of variables that should be measured in order to assess and model processes controlling water fluxes in forests. (Table 6.2)

TABLE 6.2. Variables to Measure for Assessment of Crown- and Canopy-Level Water Fluxes

Variable	Definition	Units
A_B	Basal area of trees per unit ground area	$cm^2\ m^{-2}$
A_{Fj}	Integrated area under fitted curve of radial sap flux profile	cm^2
A_S	Sapwood area of trees per unit ground area	$cm^2\ m^{-2}$
A_{Si}	AS for species i	$cm^2\ m^{-2}$
A_{Sih}	ASi for one-hectare plot	$cm^2\ m^{-2}$
A_{Sj}	Sapwood area for individual tree	$cm^2\ m^{-2}$
c_j	Distance from center of tree to centroid of fitted curve of radial sap flux profile	cm
D	Vapor pressure deficit	kPa
D_Z	Day-length-normalized vapor pressure deficit	kPa
DBH	Tree diameter at breast height	cm
E_C	Canopy transpiration	$mm\ time^{-1}$
E_{Ci}	EC for species i	$mm\ time^{-1}$
E_{Cih}	ECi for one-hectare plot	$mm\ time^{-1}$
E_S	Soil surface evaporation	$mm\ time^{-1}$
E_T	Evapotranspiration	$mm\ time^{-1}$
E_{TS}	Evapotranspiration, estimated from sap flux-scaled budget	$mm\ time^{-1}$
I_C	Canopy interception	$mm\ time^{-1}$
J_S	Sap flux density	$gH_2O\ m^{-2}\ s^{-1}$
J_{Si}	Sap flux density for species i	$gH_2O\ m^{-2}\ s^{-1}$
LAI	Leaf area index	$m^2\ m^{-2}$
LE	Latent heat flux*	$mm\ time^{-1}$
P	Precipitation	$mm\ time^{-1}$
P_T	Throughfall ($P—I_C$)	$mm\ time^{-1}$

TABLE 6.2. (*continued*)

Variable	Definition	Units
PAR	Photosynthetically active radiation	$\mu mol\ m^{-2}\ s^{-1}$
RH	Relative humidity	percent
SLA	Specific leaf area	$cm^2\ g^{-1}$
T_A	Air temperature	°C
T_B	Bark thickness	mm
T_{SW}	Sapwood thickness	cm
V_j	Volume of a rotated geometric solid	cm^3
ΔT	Temperature difference between heated and unheated sap flux probes	mV
ΔT_{max}	Maximum daily ΔT	mV
θ	Volumetric soil moisture content	m3 m^{-3}

* LE is commonly expressed in terms of Wm^{-2} but can be converted to units of mm by considering the latent heat of vaporization and the density of air for studies of the water balance.

SOURCE: From Oishi et al. (2008), with permission from Elsevier.

INTERCEPTION OF AIRFLOW

Differences in wind speed measured on windward and leeward sides of the canopy represent the effect of interception of airflow by the canopy. Dry deposition of nutrients and particles on canopy surfaces by air flowing through the canopy can be measured by rinsing foliage and measuring nutrient concentrations in the rinse water (Kimura et al. 2009) or filtering materials from airflow (Lovett and Lindberg 1993; Yokelson et al. 2007), for example, by using citric acid- and sodium carbonate-coated filters to absorb NH_3 and HNO_3, respectively, then injecting samples dissolved in deionized water through an ion chromatograph with spectrophotometric detection at 210 nanometers (Tarney et al. 2001). An airborne Fourier transform infrared spectrometer (AFTIR) also can be used to monitor transport and deposition of atmospheric chemicals and particles on canopy surfaces (Yokelson et al. 2007).

Modeling techniques have been devised to explore the effects of canopy structure on pathways of airflow and deposition of atmospheric particulates and nutrients in the canopy (Finnigan 2000; Finnigan et al. 2009; Mihailovic et al. 2009).

Eddy turbulence increases as the complexity of canopy structure increases. Turbulence affects important functions such as patterns of deposition of airborne nutrients and particulates on canopy surfaces (Tarnay et al. 2001) and dispersion patterns of biochemical signals that communicate plant or animal condition and availability of mates or prey (Mafra-Neto and Cardé 1995; Murlis et al. 1992; Visser 1986). Airflow can be measured as the direction and pathway followed by smoke or fine dyes released from a point source. Early models of canopy effect on airflow demonstrated that canopy openings caused local warming and the convective rising of air, which disrupt the laminar flow of air and create turbulence (Mafra-Neto and Cardé 1995; Murlis et al. 1992; Visser 1986). More recent studies have demonstrated that airflow is nearly linear through the canopy at low wind speeds, and the aerosol path appears as a relatively symmetrical plume oriented downwind but becomes increasingly nonlinear at higher wind speeds as deflection of airflow around barriers, including trees, generates eddies and cross-currents (Mafra-Neto and Cardé 1995; Murlis et al. 1992). Furthermore, gaps and clearings create turbulence and back flow near the edge, depending on foliage density and wind speed (Cassiani et al. 2008). Disruption in the integrity of the aerosol plume affects the ability of animals to detect the sources of attractive odors.

Figure 6.3.
The Free Air CO2 Enrichment (FACE) site in loblolly pine, *Pinus taeda,* stand at Duke Forest in North Carolina.
From Hendry et al. (1999) with permission from G. R. Hendrey and John Wiley and Sons.

EXPERIMENTAL MANIPULATION

The difficulty of manipulating canopy structure has limited experimental testing of hypotheses concerning canopy interactions with atmospheric conditions. However, several approaches have been taken to manipulate canopy effects, and deforestation of large areas has provided opportunities for the measurement of changes in atmospheric conditions and regional and global climates, resulting from canopy loss.

Free-air CO_2 enrichment (FACE) experiments have been established to evaluate the effects of increased atmospheric CO_2 on carbon exchange between canopy and atmosphere in a number of regions (Fig. 6.3), some involving mature forest trees (e.g., Crous and Ellsworth 2004; Hendry and Kimball 1994; Hendry et al. 1999; Herrick and Thomas 2003; Körner et al. 2005; Leuzinger and Körner 2007), although tropical forests are poorly represented (Stork 2007). A network of pumps and plastic tubing is used to generate increased atmospheric CO_2 concentration in treated plots (Fig. 6.4; Hendry et al. 1999). Measured variables include the canopy conditions and processes described previously (Crous and Ellsworth 2004; Hättenschwiler and Schafellner 2004; Hendry et al. 1999; Herrick and Thomas 2003; Knepp et al. 2005; Leuzinger and Körner 2007).

Removing portions of the canopy (Richardson et al. 2010; Shiels et al. 2010) or cutting individual trees or groups of trees (North et al. 2007; Schowalter et al. 2005) can create contrasts in canopy cover or structure as a means of evaluating effects of disturbance or land-use change on canopy–atmosphere interactions. Such techniques are labor-intensive and hazardous, and experimental treatments are frequently compromised by operational necessities. For example, the random selection of plot location and random assignment of treatments among plots is often impractical, given issues of site or canopy accessibility, protection of sensitive locations or species, and so on. Therefore, a more common approach has been to compare the changes in variables among unreplicated "treated" and untreated sites following operational harvest, experimental disturbance, or other manipulation. Extensive pretreatment measurement is necessary to establish essential similarity among unpaired plots and support statistical analyses of posttreatment changes (e.g., Davidson et al. 2008; Nepstad et al. 2007). Otherwise, such studies are compromised by pseudoreplication (Hurlbert 1984). Alternatively, related studies can be combined in meta-analyses, increasing independent replication to allow for robust statistical analyses that identify general trends (e.g., increased albedo and regional temperature and reduced precipitation following widespread deforestation; Gash and Shuttleworth 1991; Jactel and Brockerhoff 2007; Meher-Homji 1991).

Although manipulation of canopy organisms is more difficult, especially in mature forests (see Chapter 5), some studies have manipulated species abundances on experimental trees and evaluated effects of foliage loss on canopy temperature, precipitation interception, or carbon flux (Classen et al. 2005; Schowalter et al. 1991; Seastedt et al. 1983). Classen et al. (2005) manipulated abundances of piñon needle scale, *Matsucoccus acalyptus,* and a stem-boring moth, *Dioryctria albovittella,* by adding or removing insects from experimental trees and demonstrated that herbivory reduced foliage density enough to reduce crown interception of precipitation and affect micro-climate around treated trees.

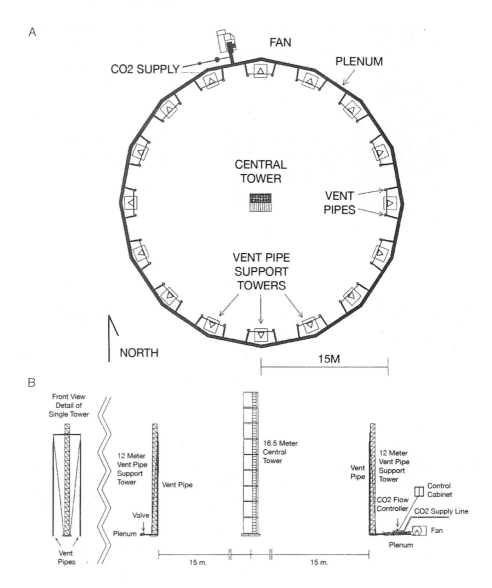

FIGURE 6.4.
Diagram of the Duke Forest FACE facility: (A) a vertical section of the FACE ring, showing the location of the plenum, towers, thirty-two vertical vent pipes located on either side of the sixteen peripheral support towers (small triangles), and fan assembly, and (B) a horizontal section through the FACE ring showing positioning of the vertical vent pipes suspended from the peripheral towers. From Hendry et al. (1999) with permission from John Wiley and Sons.

SUMMARY

Forest canopies substantially affect biotic interaction with the atmosphere. Canopy plants emit a variety of volatile organic compounds, especially isoprene and monoterpenes, that affect atmospheric redox potential and photochemical reactions. Isoprene also indirectly influences global warming through its influence on methane longevity in the atmosphere. Canopy shading of the forest floor and evapotranspiration, which promotes cloud formation and local precipitation, maintain cooler, moister conditions in forests than occur in surrounding nonforested areas. Canopies also can provide some buffering capacity against disturbances, such as storms, fire, flooding, and drought. Standard meteorological instrumentation placed above, within, and below the canopy or in nonforested sites can be used to measure the canopy's effect on these climate variables.

Forest canopies represent the biological engine that controls carbon exchange with the atmosphere through the processes of photosynthesis and respiration. These processes are fundamental to net primary production and carbon storage in forest biomass. Canopies also control fluxes of water through interception of precipitation and evapotranspiration. Finally, canopies intercept airflow and generate turbulent fluxes that influence advective fluxes and interception of particles and aerosols from the airstream. Methods for measuring these processes include eddy covariance instrumentation for carbon and water fluxes, collection of water above and below the canopy to measure interception, and rinsing foliage and filtering airflow to measure dry and wet deposition of particulates and aerosols.

Experimental evaluation of these canopy–atmosphere interactions has utilized creative approaches. FACE technology has permitted the evaluation of the effects of increased atmospheric CO_2 on carbon exchange, water fluxes, and so on. A few studies have manipulated water availability or canopy cover to measure their effects on canopy–atmosphere interaction. Finally, herbivory can affect canopy conditions and capacity for interaction with the atmosphere. Manipulation of herbivory has permitted the evaluation of its effects on water balances and climate variables.

SUGGESTED READING

Finnigan, J. J., R. H. Shaw and E. G. Patton. 2009. Turbulence structure above a vegetation canopy. *Journal of Fluid Mechanics* 637: 387–424.

Funk, J. L., and M. T. Lerdau. 2004. Photosynthesis in forest canopies. In *Forest canopies*, 2nd edition, ed. M. D. Lowman and H. B. Rinker, 335–58. San Diego, CA: Elsevier/Academic Press.

Hendry, G. R., D. S. Ellsworth, K. F. Lewin, and J. Nagy. 1999. A free-air enrichment system for exposing tall forest vegetation to elevated atmospheric CO_2. *Global Change Biology* 5: 293–309.

Körner, C., R. Asshoff, O. Bignucolo, S. Hättenschwiler, S. G. Keel, S. Peláez-Riedl, S. Pepin,

R. T. W. Siegwolf, and G. Zotz. 2005. Carbon flux and growth in mature deciduous forest trees exposed to elevated CO_2. *Science* 309: 1360–62.

Massman, W., X. Lee, and B. E. Law, eds. 2004. *Handbook of micrometeorology. A guide for surface flux measurements and analysis.* Boston: Kluwer Academic.

Oishi, A. C., R. Oren, and P. C. Stoy. 2008. Estimating components of forest evapotranspiration: A footprint approach for scaling sap flux measurements. *Agricultural and Forest Meteorology* 148: 1719–32.

Pressley, S., B. Lamb, H. Westberg, and C. Vogel. 2006. Relationships among canopy level energy fluxes and isoprene flux derived from long-term, seasonal eddy covariance measurements over a hardwood forest. *Agricultural and Forest Meteorology* 136: 188–202.

7

MEASURING CANOPY–FOREST
FLOOR INTERACTIONS

Forest canopies significantly affect, and are affected by, conditions on the forest floor. The canopy intercepts light, water, and airflow, modifying forest floor temperature and penetration of water and nutrients to the forest floor, and thereby reducing leaching, erosion, and export of sediment and nutrients from the forest. Canopy materials are exchanged with the geosphere through uptake of water and nutrients from the soil, translocation of carbohydrates to the rhizosphere, and turnover of plant and animal tissues and nutrients to the forest floor via throughfall and litterfall (Pendall et al. 2004; Prescott 2002). Exogenous factors, such as environmental changes and disturbances, and endogenous factors, such as herbivory and decomposition, influence canopy cover and the exchange of materials between the canopy and forest floor (Fonte and Schowalter 2004b; Prescott 2002).

These processes can be measured best by means of data collected above, within, and below the canopy to measure differences attributable to canopy effects (Carter and Knapp 2001; Foley et al. 2003; Fonte and Schowalter 2005; Frost and Hunter 2004; Frost and Hunter 2007; Frost and Hunter 2008; Reynolds and Hunter 2001; Schowalter et al. 1991; Seastedt et al. 1983). As for canopy–atmosphere interactions (Chapter 6), adequate representation of canopy–forest floor interaction, as affected by climate, disturbance, or herbivory, requires multidisciplinary approaches and measurement of multiple factors. This chapter focuses on canopy interactions with forest floor conditions; the exchange of water, biomass, and nutrients; and methods for experimental manipulation.

CANOPY EFFECTS ON FOREST FLOOR CONDITIONS

Forest canopies affect forest floor conditions through shading, moderation of precipitation and airflow, and turnover of biomass and nutrients. In turn, temperature and moisture at the forest floor affect water availability for plant growth, decomposition, and soil respiration. This section focuses on canopy effects on forest floor temperature and moisture conditions and on water, carbon, and nutrient fluxes.

Forest canopies intercept sunlight, reflecting heat and shading the forest floor, thereby reducing albedo and cooling the canopy and forest floor (Foley et al. 2003; Gash and Shuttleworth 1991; Janssen et al. 2008). This insulating effect increases with canopy cover and canopy depth. Temperature at the forest floor under dense, complex canopies typically remains 2–4°C cooler than the top of the canopy, with the difference between canopy surface and forest floor reaching a maximum of 10–12°C at midday, when ambient temperature peaks; this difference disappears at night (Foley et al. 2003; Madigosky 2004; Parker 1995). Litter falling from the canopy (as discussed later) further insulates the soil surface from temperature extremes. As a result of this canopy cooling effect, temperatures at the forest floor are relatively constant diurnally and seasonally, providing stable conditions for a variety of organisms and processes.

Canopy opening as a result of disturbance or herbivory disrupts this shading effect. Solar exposure can raise soil surface temperatures to 45°C during midday (Seastedt and Crossley 1981), creating adverse conditions for many forest floor organisms that control decomposition and soil fertility (Amaranthus and Perry 1987) and increasing evaporative loss of water. Furthermore, the effects of such soil warming can extend as much as 200 meters into undisturbed forest, creating a horizontal gradient in forest floor temperature (Chen et al. 1995). Loss of vegetation cover can initiate a positive feedback between evaporation and reduced precipitation that leads to further vegetation loss (Janssen et al. 2008).

Soil moisture is a function of precipitation reaching the soil (throughfall), uptake by vegetation, and evaporation. Throughfall is determined by the amount of precipitation intercepted by the canopy and reflects canopy cover and surface area. Vegetation uptake is determined by water requirement and rate of evapotranspiration by the canopy. Finally, evaporation from the forest floor is determined by relative humidity.

Relative humidity at the forest floor is typically higher than at the top of the canopy, as a result of lower temperature (as discussed previously) and airflow, with the gradient particularly pronounced at midday (Madigosky 2004; Parker 1995). High relative humidity and low airflow minimize direct evaporation of soil moisture. However, when the canopy is opened as a result of disturbance or herbivory, soil exposure and warming increase the rate of evaporation and can lead to soil desiccation. Counteracting this trend is the reduced interception of precipitation and uptake of soil water by the opened canopy, increasing soil moisture. Depending on drainage conditions, increased soil water could increase leaching and export or result in flooding of the site. The relative amounts of precipitation input to soil and evaporation or drainage from soil determine soil moisture under these altered conditions.

WATER AND NUTRIENT FLUXES

Canopy structure determines both the demand for water by the canopy and the rates of precipitation interception (see Chapter 6) and throughfall to the forest floor. These processes govern water balance and cycling by the forest. In addition, canopy biomass and nutrients are transferred to the forest floor via throughfall enrichment.

Plants require water and nutrients for photosynthesis and canopy growth. Some water is absorbed through foliage surfaces, but most water and nutrients for photosynthesis and canopy growth must be transported upward from the soil, against gravity. This process is driven, in large part, by capillary action aided by evapotranspiration, which maintains the negative pressure in xylem necessary to draw water from the soil. Path length resistance limits the height to which water can be drawn through capillaries, restricting maximum plant height to 120–30 meters (Koch et al. 2004). If soil water becomes limiting, xylem cells cavitate, and the plant exhibits symptoms of drought stress (Mattson and Haack 1987; Trumble et al. 1993). Some bark beetles detect and use cell cavitation as a cue to inhabit water-stressed plants that are less able to produce defensive compounds and thereby become more suitable hosts (Mattson and Haack 1987). If water limitation is severe, portions of the canopy die, leading to lateral branching—the reiteration of trunks arising from the main trunk (Sillett and Van Pelt 2007) and/or development of platforms that contribute unique habitats for various organisms.

The canopy also intercepts and modifies precipitation, determining the evapotranspiration rate (see Chapter 6), throughfall (water dripping from the canopy to the forest floor) and stemflow (water flowing down boles) chemistry, droplet impact on the forest floor, and erosion (Brauman et al. 2010; Foley et al. 2003; Pypker et al. 2005). As precipitation percolates through the canopy, much remains on foliage and other surfaces, reducing the volume reaching the forest floor. Furthermore, water dripping from lower canopy surfaces has much lower energy and continues over a longer time period, compared to unmodified precipitation, thereby reducing its impact on the forest floor and ability to dislodge and move soil or other materials (Meher-Homji 1991; Ruangpanit

1985). Interception rate increases with increasing canopy surface area and decreasing precipitation volume (Brauman et al. 2010).

Throughfall and stemflow show chemical enhancement, relative to raw precipitation, as a result of acquisition of nutrients from material adhered to or leached from foliage and branches during downward flow from the canopy. Foliage fragmentation resulting from herbivory or storm damage increases leaching from open edges of leaves (Kimmins 1972; Seastedt et al. 1983; Schowalter et al. 1991). Increased nutrient content of throughfall increases flux of nutrients from canopy to forest floor. Water reaching the forest floor in excess of soil storage capacity leaches into streams and is exported from the forest.

FLUXES TO THE FOREST FLOOR AND RHIZOSPHERE

Biomass and nutrients are transferred from the canopy to the forest floor through several pathways. One is translocation of canopy photosynthates (carbohydrates) to roots, which use them for growth but also exude substantial quantities into the rhizosphere to support microbial associates that contribute to water and nutrient uptake. A second pathway is litterfall, by which plant and animal tissues fall to the forest floor as litter. Dead plant and animal biomass is decomposed and the carbon and nutrients are released for use by soil/litter organisms or uptake by plants. Some soluble nutrients are leached through the soil and exported from the forest ecosystem.

Carbohydrates produced by photosynthesis in the canopy support metabolic activity throughout the plant. Carbohydrates move downward through the phloem and are stored in woody tissues and roots. Allocation of net primary production to below-ground plant parts is often 50 percent or more in forests (Coleman et al. 2004). Furthermore, 20–50 percent more carbon enters the rhizosphere from root exudates and exfoliates than is measured in root biomass at the end of the growing season (Coleman et al. 2004). Root exudates support a variety of associated organisms, particularly nitrogen-fixing bacteria and mycorrhizal fungi, that are critical to adequate uptake of water and nutrients by roots (Coleman et al. 2004). Exudates also contribute to soil aggregate formation, a process that increases soil nutrient retention (Coleman et al. 2004).

Litterfall consists of a variety of canopy materials falling to the forest floor. These include foliage, reproductive structures and woody debris, and animal feces and tissues. These materials vary in their contributions of biomass and nutrients to the forest floor. Fresh foliage has higher nutrient concentrations than does senescent foliage (Fonte and Schowalter 2004a), due to resorption by the plant prior to abscission (Marschner 1995). Concentrations of nutrients in woody material are very low. Animal feces and tissues are high in nutrients and are known to stimulate decomposition and mineralization (Fonte and Schowalter 2005; Frost and Hunter 2004; Frost and Hunter 2007; Frost and Hunter 2008; Schowalter and Crossley 1983; Seastedt and Tate 1981). Litter provides habitat for a variety of forest organisms.

Canopy conditions affect rates of decomposition and mineralization of litter on the forest floor and soil respiration by forest floor organisms. The degree of canopy shading and amount of throughfall govern litter temperature and moisture—two factors that control decomposition and respiration rates (Meentemeyer 1978; Prescott 2002; Seastedt 1984; Whitford et al. 1981). Decomposition rate also is a function of litter quality, as determined by tree species and litter material (e.g., foliage versus wood; Fonte and Schowalter 2004a, b; Prescott 2002).

EFFECTS OF CANOPY ASSOCIATES

Canopy organisms can substantially affect canopy–forest floor interactions. A number of organisms affect fluxes of materials to the forest floor. Epiphytes added 140 kilograms per tree in a temperate rain forest in Chile (Díaz et al. 2010) and added additional mass when filled with water. Breakage of overweighted branches during storms is common.

A number of insects have life cycles that are divided between canopy and forest floor habitats. Some folivores feed on canopy resources as immatures and pupate on the forest floor (e.g., Miller and Wagner 1984). Others feed on below-ground tissues as immatures but emerge and affect canopy structure as adults (e.g., cicadas, which can cause substantial twig and foliage loss during oviposition in twigs).

Insect outbreaks add substantial amounts of fecal material and green foliage fragments, as well as nutrient-enhanced throughfall (Grace 1986; Frost and Hunter 2004; Frost and Hunter 2007; Frost and Hunter 2008; Hollinger 1986). Roosting birds also add weight that can break branches and feces that enrich soils below rookeries. In addition, some forest floor residents (e.g., elephants and giraffes) can reach and substantially influence canopy structure and function up to 4–5 meters in height, and many forest floor predators forage into forest canopies.

Tree-hole mosquitoes may seek hosts on the forest floor as well as in the canopy. Furthermore, deforestation can increase standing water, hence mosquito habitat, in open areas (as described previously) and thereby increase the incidence of mosquito activity and disease transmission (Vittor et al. 2006).

MEASUREMENT OF CANOPY–FOREST FLOOR INTERACTION

A number of methods are available for measuring canopy effects on the forest floor and fluxes of material between the canopy and forest floor (Table 7.1). One's selection of variables and methods depends on research objectives.

FOREST FLOOR TEMPERATURE AND MOISTURE

Forest floor temperature is measured most conveniently using themisters or thermocouples placed on the soil/litter surface and/or at various depths, depending on research

TABLE 7.1.   Methods for Measuring Canopy–Forest Floor Interaction

Variable	Instrument	Method	Frequency
Soil temperature	Thermistors	Data logger	Hourly/integrate daily
Soil moisture	Thermistors	Data logger	Hourly/integrate daily
Interception	Bottle	Mass balance	After rainfall event
Water transport	Sap flux probe	Flux rate	Monthly, day/night
	Pressure bomb	Flux rate	Monthly, day/night
	Stable isotope	Flux rate	Monthly, day/night
Throughfall	Bottle	Mass balance	After rainfall event
Litterfall	Litterfall collector	Mass balance	Monthly
Frassfall	Litterfall collector	Mass balance	Daily, rain-free period
Decomposition	Litterbag/tethered	Mass balance	Variable
Nutrient concentrations	Auto-analyzer/AA	Spectrophotometry	Variable

objectives. These devises are attached to a data logger that records temperatures at designated time intervals and provides daily means, maxima, and minima. Changes in forest floor temperature resulting from differences or changes in canopy opening can be measured by comparison of means of appropriately replicated canopy treatments.

Soil moisture can be measured gravimetrically or by time domain reflectometry, using sensors inserted into the soil at various depths and connected to a data logger (Irvine et al. 2002; Oishi et al. 2008). Site-specific calibration of sensor data using gravimetric analysis is required. Changes in soil moisture resulting from differences or changes in canopy cover can be measured by comparison of means of appropriately replicated canopy treatments.

WATER AND NUTRIENT TRANSPORT INTO THE CANOPY

Several methods are available for measurement of plant water uptake and use by the canopy (Wullschleger et al. 1998). Pressure bombs (e.g., Scholander et al. 1965) placed on the end of a cut petiole or twig can be used to measure xylem water potential (e.g., Irvine et al. 2002; Kolb et al. 1999). Measurements generally are made just before dawn to avoid confounding measurement with increasing water stress during the day. Measurements can be taken at various canopy levels for comparison. More negative readings indicate increasing moisture deficit. Effects of drought or other water limitation can be detected as increased pressure values. Cell cavitation during extreme moisture limitation can be detected by sensitive audio equipment (Mattson and Haack 1987).

The measurement of stomatal conductance provides an estimate of plant water uptake, as well as evapotranspiration (Irvine et al. 2002; Wullschleger et al. 1998; see Chapter 6). However, extrapolation of stomatal conductance at the leaf scale to water use by the plant is difficult because the higher humidity in the boundary layer of leaves decouples vapor pressure at the leaf surface from that of the air, altering the rate of leaf and canopy transpiration (Wullschleger et al. 1998). Water stress causes plants to reduce stomatal conductance.

Rates of xylem sap flow can be measured as sap flux (Granier 1987; Irvine et al. 2002). Two probes are inserted radially into the outer xylem and 10 centimeters apart vertically (Granier 1987). The lower (heater) probe is wrapped tightly with a heating element consisting of a coil of insulated constantan wire supplied with a constant voltage to provide a power dissipation of 0.2 watts. This probe is coated with a thermally conductive paste and placed in an aluminum tube inserted previously into a radially orientated drill hole into the xylem. The upper (reference) probe is inserted into a similarly orientated drill hole situated 10 centimeters above the lower probe. Both probes are attached to thermocouples connected in opposition. The sap flux is influenced by the temperature difference between the two probes, stored on a data logger. Sap flow velocities (F_s) along a radius (m sec^{-1}) are calculated from differences in temperature between the two probes as

$$F_s = 119 \times \frac{(\Delta T_m - \Delta T)^{1.23}}{\Delta T},$$

where ΔT_m and ΔT are the temperature differences between the two probes, for no flow and positive flow, respectively (Oishi et al. 2008). Total sap flow F (m^3 sec^{-1}) is

$$F = F_s S_a,$$

where S_a is the cross-sectional area (m^3) of sapwood at the heating probe (Granier 1987; Oishi et al. 2008).

Stable isotopes, such as deuterium, offer a method for measuring water uptake and movement within trees that is relatively free of the limitations associated with calculation of water uptake from energy balance, vapor pressure, or heat transfer (Wullschleger et al. 1998). Labeled water can be provided as augmented soil water or introduced into drilled holes in the bole. Detectors above the point of input are used to measure the rate of movement directly.

Infrared gas analyzers (IRGA) and eddy-covariance techniques are becoming widely used to measure fluxes of water between the forest floor, various canopy levels, and the atmosphere (see Chapter 6). Eddy covariance measures of evapotranspiration provide estimates of water uptake by trees and other vegetation.

Recently, remote-sensing techniques have been developed to measure effects of various plant stressors on a variety of plant species, including conifers and deciduous trees (Carter and Knapp 2001), as this may affect foliage area and water fluxes. A foliage sample is placed on a black platform under a high-intensity tungsten lamp. Reflectance is measured at 1-nanometer wavelength intervals using a spectroradiometer attached by fiber-optic cable to a telescope/microscope body. Radiance reflected from the foliage sample is divided by radiance reflected from a white reference to compute reflectance in units of percentage. Data are recorded at 1-nanometer intervals throughout the 400–850 nanometer range. In all cases, the maximum difference in reflectance between stressed and control trees was expressed as a reflectance increase at wavelengths near 700 nanometers. This optical response can be explained by a general tendency for stress to reduce chlorophyll concentrations in foliage. Nansen et al. (2010) used a hyperspectral camera with a wavelength range of 405–907 nanometers to evaluate the effects of severe, moderate, or no drought stress or spider mite (*Tetranychus uricae*) infestation in maize. They found a particularly strong response to drought stress at 706 nanometers, with no significant response to spider mite infestation, but a significant response to spider mites, as well as drought stress, at 440 nanometers.

Nutrient uptake from the soil occurs primarily as soluble nutrients entering root cells with water. Rates of nutrient movement upward in trees can be measured using stable isotopes (e.g., ^{15}N) added to soil and monitored as they move into and through the plant (Fig. 7.1; e.g., Coleman et al. 2004; Frost and Hunter 2004; Frost and Hunter 2007).

WATER AND NUTRIENT TRANSPORT TO THE FOREST FLOOR

Precipitation volume and duration are measured using standard-weighing rain gauges placed above the canopy or in nearby gaps or nonforested sites. Tipping-bucket rain gauges use a funnel to collect and channel precipitation into a small container balanced on a lever. When a sufficient amount of precipitation has collected, the lever tips, dumps the collected water, and sends an electrical signal that records the volume and time interval. This type of rain gauge may undermeasure precipitation because a rain event may end before the bucket tips.

Interception volume can be measured as the difference between precipitation volume measured above the canopy and throughfall volume measured below the canopy, adjusted for water flowing down main boles (Fig. 7.2; Pypker et al. 2005; Schowalter et al. 1991). Similarly, interception at various canopy levels can be measured by difference in volumes from one level to the next. Measuring changes in interception resulting from canopy development, seasonal foliage loss, and herbivory or disturbances requires appropriate replication of canopies representing different developmental stages, seasons, and other factors.

Throughfall can be collected in rain gauges or other containers, often with funnels to increase the amount collected for measurement of nutrient content (Pypker et al. 2005; Schowalter 1999). The difference in nutrient concentrations between precipitation and throughfall represents enhancement during percolation through the canopy. Nutrients in throughfall and precipitation can be weighted by volumes to calculate nutrient flux for the intercepted area.

Stemflow is more difficult to measure but may be relatively unimportant to water flux measurements. Oishi et al. (2008) estimated that stand-level stemflow accounted

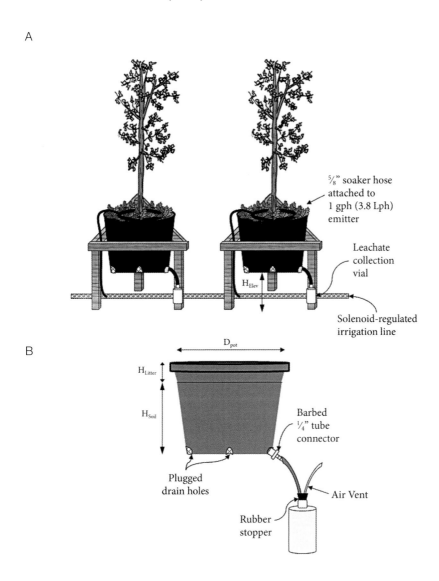

FIGURE 7.1.
Design of experiment to measure fluxes of canopy C and N in defoliated or nondefoliated red oak (*Quercus rubra*) saplings to and from the soil as affected by herbivore (*Malacosoma americanum*) feces. Fluxes from the soil were measured using stable isotopes. Fluxes into the soil were measured with ion exchange resin bags, and nutrients exported from the pots. From Frost and Hunter (2004) with permission from the Ecological Society of America.

Figure 7.2.
Experimental evaluation of manipulated herbivore density on water and nutrient fluxes in young conifer forests. The metal pan (with a thirty-six-degree angle positioned against the main bole and covered beyond the canopy perimeter) was designed to collect 10 percent of material (throughfall, greenfall, carcasses, and frass) falling from/through the canopy. Identical collectors in open areas were used for comparison to separate atmospheric and treatment inputs. Litterbags under the tree were used to evaluate effects of herbivore inputs on decomposition.
From Schowalter et al. (1991) with permission from Elsevier.

for 1 percent of annual precipitation. Collectors can be secured at various heights above the ground on tree boles to measure the volume of water flowing down (Schowalter et al. 1991).

Impacts of precipitation and/or throughfall can be measured as displacement and transport of soil and litter (erosion) by water reaching the forest floor using sediment collectors. These are shallow pans installed flush with the soil surface to collect sediment displaced by raindrop impact.

Nutrient translocation to roots and rhizosphere is most easily measured using stable isotopes, such as ^{14}C and ^{15}N (Coleman et al. 2004; Holland et al. 1996). Labeled photosynthate can be tracked from the source in foliage through the plant and into roots and mycorrhizae or other soil microorganisms. Fluxes of soluble nutrients from the canopy into soil are measured using ion exchange resin bags or lysimeters (Schowalter et al. 2012). Resin bags are constructed with nylon stockings, and each bag is partitioned to hold anion resin beads on one side and cation resin beads on the other. Cation beads absorb anions (such as NH_3, NO_3, and PO_4), and anion resin beads absorb cations (such as K, Ca, and Mg). Stockings cut into strips are formed into pouches using hot glue. Each pouch is filled with a weighed amount of the designated beads. All materials should be handled with latex gloves to avoid contamination. Resin bags can be placed either at the litter/mineral soil interface to measure fluxes entering the soil or at designated depths in the soil to measure flux through the soil. Individual bags are collected at designated intervals and dried. Soil adhering to the bags should be removed gently prior to nutrient analysis.

LITTERFALL

Litterfall collectors generally are constructed as baskets intercepting a known horizontal area and capable of draining precipitation/throughfall (Fig. 7.3). Small containers such as 1- or 5-gallon buckets have a tendency to underestimate total litterfall, since they frequently tip over and also deflect large branch fall. Cubic-meter mesh litter traps on stakes above ground are not only relatively accurate because they retain larger litterfall but also easy units for extrapolation to whole forest litter (reviewed in Lowman 1985). In addition to the challenges of large branch fall, feces (especially insect feces) are easily disintegrated, and nutrients are lost as a result of rain (Mizutani and Hijii 2001; Southwood 1978). Therefore insect frass collection is often restricted to short, rain-free periods (e.g.,

Figure 7.3.
Litterfall collector illustrating a one-meter trap.

Fonte and Schowalter 2004a), although Mizutani and Hijii (2001) described a method for correcting frass reduction due to rain. Collected litter components (including fecal material) can be separated, dried, and analyzed for nutrient concentration using standard spectrophotometric techniques (Greenberg et al. 1992). Nutrient concentrations weighted by litter mass represent flux rate.

DECOMPOSITION AND SOIL RESPIRATION RATES

Decomposition can be measured by several methods differing in requirements for time and materials. Measuring the mass of litterfall and litter standing crop per unit area provides a ratio that represents the decay constant, k, when litter mass is constant (Olson 1963). This method suffers from the difficulty of separating litter from underlying soil for accurate measurement of litter standing crop. Respiration rate from decomposing litter can provide an estimate of decay rate, but separating respiration of litter from that of underlying root and rhizosphere respiration is difficult. An inexpensive method involves a PVC chamber containing a weighed amount of soda lime or a solution of NaOH sealed over litter for twenty-four hours, measuring respiration as the rate of CO_2 absorption (Fig. 7.4). Efflux of CO_2 is measured as the increase in mass of soda lime or the volume of acid neutralized by NaOH (Edwards 1982). Stable

FIGURE 7.4.
PVC pipe respiration chamber into which an open jar of weighed soda lime or NaOH solution is placed and the chamber capped for twenty-four hours. In this example, respiration was measured as increased weight due to CO_2 absorption by soda lime or NaOH. Tents were used to compare respiration when litter was exposed to, or protected from, throughfall.

isotopes (e.g., ^{13}C, ^{14}C, and ^{15}N) can be used to measure loss and transfer of important nutrients (e.g., Šantrůčková et al. 2000). Infrared gas analysis is a convenient method for measuring respiration (Nakadai et al. 1993).

Decomposition most commonly is measured using tethered litter or mesh litterbags containing a measured weight of litter. Tethered litter is most accessible to large (e.g., millipedes) and small (e.g., oribatid mites and collembolans) detritivores that control initial fragmentation and inoculation of microbes. Selected mesh size of litterbags requires compromise between allowing maximum access to detritivores of various sizes and minimizing effects on litter moisture versus retaining litter fragments (Coleman et al. 2004). The litter can be composed of senescent foliage or other components of individual tree species or a representative mix of litter materials (e.g., Fonte and Schowalter 2004a), depending on research objectives. Subsamples of litterbags should be retained and dried, if necessary, to calculate fresh weight–dry weight ratios and for analysis of initial nutrient concentrations, as desired.

To measure effects of canopy structure, disturbance, or herbivory on litter decomposition, plots can be established at locations under replicated canopy treatments (e.g., Seastedt et al. 1983; Schowalter et al. 1991). Multiple litter samples are placed on the litter surface in the designated areas and gradually become incorporated into the litter as fresh litter falls. Individual litter samples are collected at designated time intervals, typically two, four, eight, twelve, twenty-four, or forty-eight weeks for litter with rapid decay rates (e.g., foliage) or longer intervals for litter with slow decay rates (e.g., woody material)

Remaining litter is weighed and decay rate is calculated using a single- or double-component negative exponential model (Olson 1963):

$$N_t = S_o e^{-kt} + L_o e^{-kt},$$

where N_t is mass at time t; S_o and L_o are masses in short- and long-term components, respectively; and k's are the decay constants. Alternatively, litter material labeled with stable isotopes or rare Earth elements can be used to estimate rate of litter decay. Decay rates and/or loss of nutrients from litter can be compared among litter species or canopy treatments to evaluate the effects of canopy conditions on this key ecosystem process.

Several methods are available to measure soil respiration. The first is the CO_2-absorption method (discussed previously). Respiration chambers can be constructed from PVC piping with an airtight lid on one end (Fig. 7.4; Edwards 1982). The open end is inserted into the soil to a premarked depth to represent respiration within particular volumes of soil and chamber. An open container with a measured mass of soda lime is placed in the chamber, and the lid is sealed in place with grease or petroleum jelly. After twenty-four hours, the soda lime container is capped and reweighed. The change in mass represents absorbed CO_2, reflecting the rate of soil respiration. Alternatively, an infrared gas analyzer can be used to instantaneously measure CO_2 flux from the soil (Davidson et al. 2008; Nakadai et al. 1993). Eddy covariance methods also have been used to measure forest floor respiration (see Chapter 6).

EXPERIMENTAL MANIPULATION

The difficulty of manipulating forest canopy structure has limited experimental testing of many hypotheses related to canopy effects on forest floor conditions. In some cases, large-scale unreplicated treatments can be validated by careful pretreatment measurements that establish the similarity of two plots prior to the treatment of one (Davidson et al. 2008; Hurlbert 1984). This permits the comparison of divergent changes following treatment. For example, Davidson et al. (2008) measured soil gas emissions from two 1-hectare plots in a mature tropical forest for 18 months prior to experimental drought treatment. They then suspended photosynthetically-active radiation (PAR)-transmitting plastic sheeting above the forest floor, using wooden frames, in one plot during the wet season to simulate drought and compared changes in CO_2, CH_4, and N_2O emissions from the forest floor between the two plots during five years of treatment followed by one year of recovery. Other experimental approaches also have been taken to evaluate canopy effects on soil temperature, moisture, and fertility.

Canopy effects on soil temperature can be manipulated using buried electrical grids in order to evaluate the effects of soil warming on litter decomposition and soil respiration (Melillo et al. 2002). Similarly, canopy effects on soil moisture can be manipulated by means of irrigation to increase precipitation/throughfall input to the forest floor or sheltering treatments that reduce precipitation/throughfall input to the forest floor (e.g., Davidson et al. 2008; Irvine et al. 2005). The techniques to measure decomposition and respiration described previously can be used to compare decomposition and soil respiration among replicated plots.

Free air CO_2 enrichment (FACE) experiments (see Chapter 6; Fig. 6.3) have provided the opportunity to study the effects of increased atmospheric CO_2 on a variety of ecosystem processes, including herbivory and carbon and nutrient fluxes to the forest floor (Hättenschwiler and Schafellner 2004; Hendry and Kimball 1994; Hendry et al. 1999; Knepp et al. 2005; Lindroth et al. 2001). Such facilities are expensive to establish, and their use for experiments requires access to a FACE site.

Canopy cover and depth have been manipulated in some studies by removing portions of the canopy (Richardson et al. 2010; Shiels et al. 2010) or cutting individual trees or groups of trees (Fig. 4.3; North et al. 2007; Schowalter et al. 2005) to evaluate effects of changes in canopy cover or structure, such as those resulting from disturbances, on forest floor conditions and processes. In a particularly ambitious project, climbers trimmed tropical forest canopies to where bole or branch diameters reached 10 centimeters, to simulate hurricane disturbance to the canopy in replicated plots (Richardson et al. 2010; Shiels et al. 2010). In a subset of trimmed plots, trimmed litter was transferred to an uncut plot to simulate the effect of the resulting pulse of litterfall without the associated canopy opening. Canopy opening invariably increases albedo, heating the forest floor during the day and increasing temperature variation (Foley et al. 2003; Gash and Shuttleworth 1991; Janssen et al. 2008), thereby altering rates of gas fluxes, decomposition, and other ecosystem processes.

Although more difficult to accomplish in mature forests, manipulation of herbivory among replicated trees or plots can be used to demonstrate effects on precipitation interception, throughfall volume and chemistry, and litter decomposition and soil respiration (Boxes 7.1 and 7.2; Chapman et al. 2003; Classen et al. 2005; Fonte and Schowalter 2005; Frost and Hunter 2004; Frost and Hunter 2007; Frost and Hunter 2008; Kimmins 1972; Schowalter et al. 1991; Seastedt et al. 1983; see Chapter 6). Herbivores can be added and/or removed manually or reduced with insecticides (see Chapter 5). Addition requires a source of herbivores from surrounding vegetation or from laboratory colonies. Since some insecticides include elements of interest, especially nitrogen, selection of insecticide should address potential confounding effects on nutrient fluxes of interest. Herbivory can be simulated by foliage punching or clipping, but this does not represent all effects of herbivory, especially deposition of feces, which can have important effects on soil fertility and decomposition (Chapman et al. 2003; Fonte and Schowalter 2005; Frost and Hunter 2004; Frost and Hunter 2007; Frost and Hunter 2008; Schowalter et al. 2012). Effects of herbivory also can be simulated by augmentation of greenfall, feces deposition, and throughfall in experimental plots (Fig. 7.5; Frost and Hunter 2007; Frost and Hunter 2008; Schowalter et al. 2012). However, this method does not represent the effect of canopy opening equivalent to the simulated level of herbivory.

FIGURE 7.5.
Small plot manipulation with augmented greenfall (fresh foliage material resulting from insect herbivore feeding) to test effects on litter decomposition (note litterbags) and soil nutrient flux (using buried ion exchange resin bags). Mesh litterbags contain a weighed mass of leaf litter and are harvested at designated intervals, reweighed, and analyzed for mass and nutrients remaining.

BOX 7.1: CONFESSIONS OF A FRASS LOVER: METHODS SUMMARY FOR CANOPY–FOREST FLOOR INTERACTIONS

Chris Frost

One fundamental interaction in nature exists between plants and their herbivores, and this interaction has a significant effect on how ecosystems function. In terrestrial forests, canopy herbivores that consume leaf material are ubiquitous, and there is growing evidence that canopy herbivores influence the forest soil below them. But how? Several mechanisms may allow canopy herbivores to alter soil processes, and these mechanisms operate across a range of spatial and temporal scales. While I have studied a number of these mechanisms, one of the more obvious ways that a canopy herbivore could influence soil processes is through their feces. Herbivores defecate as they eat; such fecal material falls down to the forest floor and, in fact, can actually be heard falling when herbivore densities are high. Prior to my work, insect canopy herbivore feces ("frass") had been correlated with soil nutrient pulses; I wanted to test whether frass and soil nutrient dynamics were linked causatively. I did this with a series of experiments using an experimental field site that I designed and built, which allowed me to manipulate insect herbivory and directly measure links between insect herbivore feces and soil processes. These experiments revealed that herbivore feces mediate a fundamental relationship between forest soils and insect canopy herbivory on red oaks. This basic design, while difficult and somewhat costly (with the isotope analysis), would lend itself well to other forest species and forest ecosystem types.

In some cases, ecological studies benefit from controlled field conditions. Such was the case, or so I argued, when I established a potted field study to address the effects of canopy herbivory on soil processes. I wanted to be able to replicate field conditions in a way that also allowed me to have a large number of experimental replicates in close proximity, which is termed a common garden experiment. However, in my case I wanted to use soil and surface leaf litter from the field sites where the observations linking insect canopy herbivory and soil nutrient availability had been made. Since there was no way to clear-cut a mountain forest, and the soil at my field site was not comparable, I settled on transporting soil from the mountain. This demanded that I use pots. To make life more complicated, since I wanted to be able to analyze the water quality that leached from the pots, each pot needed to be suspended above the ground. I settled on building a series of identical pot stands that held my experimental trees. The stands were made from pressure-treated wood and designed to suspend the 7-gallon pots that held the northern red oaks (*Quercus rubra* L.). I collected leachate by sealing three of the four drain holes in the bottom of the pot and connecting the fourth to a collection bottle.

Experiments that manipulate insect herbivore activity (e.g., feeding damage) often require that herbivores be contained. In my case, I hand sewed bags made of Reemay cloth to place over the plants and enclose the herbivores. Insect herbivores are placed inside the bag, which is tied off to prevent escape. Remember that bags must be included with controls that do not have herbivores to account for any effect of the bag on the process of interest. The bag size depends on the experiment. Regardless of the size of the cage, I had to monitor the bags frequently to ensure that the herbivores were feeding, determine the rate of damage, and monitor frass production.

Initially, I wanted to test the effect of frass additions independently of the physical damage inflicted by the herbivores. I used frass produced from herbivores feeding on the trees in the common garden, which had to be added to the surface soil of some trees that had been damaged by herbivores and some trees that had not been damaged. To do this, I used a "frasspirator" to collect the frass from inside each bag, which was then pooled and divided into equal amounts to fill the same number of pots from which the frass had been collected. So the amount of frass added to each pot represented an average of what would have fallen into any of the pots that had herbivores. After the frass was added, I measured nutrients in the soil and leachate over time. Soil measurements were made by extracting soil samples with 1 M potassium chloride followed by filtration with no. 1 Whatman paper, and the filtrate was processed with an Alpkem segmented flow autoanalyzer. Nutrients in the leachate were measured directly on the Alpkem.

A subsequent experiment used stable isotopes to physically trace the flow of frass nitrogen into my experimental version of the forest floor. To do this, I enriched a set of "sacrificial" trees apart from the common garden with ^{15}N to generate labeled frass and then deposited the "labeled"

BOX 7.1, FIGURE 1. Insect herbivores like this gypsy moth (*Lymantria dispar* L.) can have large influences on how forest ecosystems function, partially due to their year-to-year variation in population densities but also as a result of how they modify the traits of the plants on which they feed.

(continued)

(continued)

frass onto the soils in the common garden. First, I added a 50-millimolar solution of 99-atom percent ^{15}N ammonium nitrate to the potting soil and allowed the sacrificial trees to incorporate the ^{15}N into their leaves. I then used insect herbivores to completely defoliate this set of trees and collected all the frass. After measuring the ^{15}N concentration in the frass using an isotope ratio mass spectrometer (IRMS), I added an amount of frass to a set of trees in the common garden. I then measured the accumulation of ^{15}N (derived from the frass) over the course of time into the soil, leachate, and tree. Natural abundance of ^{15}N is 0.3663 per million; I calculated ^{15}N-enriched additions carefully to achieve ^{15}N enrichment between 0.5 and 1 per million in the soil, leachate, and tree samples. In all cases, an IRMS was required to determine ^{15}N concentrations in the samples relative to reference samples (i.e., with no ^{15}N enrichment). Note that care must be taken when handling stable isotopes to ensure that contamination does not occur, as even a small amount of enriched stable isotope will contaminate a sample at natural abundance. In addition, ^{15}N overenrichment is a common problem for those new to stable isotope enrichment studies; such samples are a waste of money and compromise the IRMS. Be very careful during experimental design to calculate target enrichment concentrations of roughly double natural abundance.

BOX 7.2: HERBIVORE REMOVAL EXPERIMENTS

Samantha Chapman

Though herbivores most directly alter plants through their consumption of biomass, they can also alter forest structure and ecosystem processes in other important ways (Hunter 2001; Schowalter 2011). In particular, herbivores that chronically infest forests year after year can exert important influences on ecosystem dynamics. In a shrub-dominated woodland, pinyon pines resistant and susceptible to two chronic herbivores form a mosaic of tree phenotypes on the landscape. We have supplemented this "natural experiment" with long-term herbivore removal experiments (Cobb et al. 1993; Gehring and Whitham 1995). Herbivore removals have allowed us to examine how herbivores can alter tree structure, soil microclimate, and litter chemistry, thereby impacting carbon and nitrogen cycling.

Herbivore removal experiments allow us to definitively attribute altered plant and ecosystem characteristics to the focal herbivores in this ecosystem.

We annually remove scale insects, which infest juvenile pinyon pines, by removing the egg mass at the base of individual trees prior to scale emergence. We remove moth larvae, which feed on reproductively mature pinyon pines, by spraying individual trees with insecticide each year prior to moth emergence. By comparing scale susceptible trees to scale removed and scale resistant trees, we have shown that scale insects alter pinyon architecture by creating a "poodle-tail" appearance, in which all but the most recent two years of needles abscise and generate an "open" canopy. Conversely, stem-boring moth larvae generate a prostrate shrubby canopy and therefore a more "closed" tree architecture. To assess canopy changes, we counted the number of shoots in a given canopy area using a pie-wedge-shaped PVC frame. Because trees represent "islands of fertility" on the landscape, these canopy changes can have large impacts on both inputs and microclimate.

Measuring the impact of insect herbivory in these ecosystems is challenging because the layout of plants does not resemble closed-canopy forests or grasslands—ecosystems where most herbivore-assessment techniques were developed. Thus we had to develop a suite of tree-based techniques to monitor herbivore impacts on ecosystem dynamics. For example, to assess litterfall, we custom built isosceles triangle-shaped litter traps for each study tree. These litter traps accurately collect a known proportion (1/20) of total tree litterfall, thus allowing us to scale up to ground area–based litterfall rates. Litterfall collections using these traps yielded evidence that herbivores can impact both litter quantity and quality (Chapman et al. 2003; Classen et al. 2007). Above-ground litter decomposition measurements were performed using custom-made litter decomposition bags composed of a small (0.2 millimeter) polypropylene mesh on the bottom to prevent needle loss and a larger 0.8 millimeter polyester mesh on top to give microarthropod decomposers access to litter. Root litter decomposition rates were assessed using mesh bags (both sides made of 0.8 polyester mesh) that were buried vertically, 15 centimeters deep in the soil profile at the tree crown drip line, and tethered to the trunk using nylon fishing line. We examined the interaction between herbivore-altered soil microclimate and litter chemical quality by performing reciprocal litter transplants. In other words, herbivore-susceptible and herbivore-resistant litter was decomposed under both herbivore-susceptible and herbivore-resistant trees.

Herbivore-modified tree architecture can change soil microclimate (Classen et al. 2005). We examined the impacts of altered tree architecture on soil moisture using time-domain reflectometry (Trase System 1, Soil Moisture Corporation, Goleta, CA). Volumetric soil moisture was measured at three depths (0–15, 16–30, and

(continued)

(continued)

31–45 cm) at two-week intervals during the growing season. Soil temperature was measured using Stowaway XTI temperature thermistor probes (Onset Computer Corp, Bourne, MA). Probes were buried at 5-cetimeter and 15-centimeter depths, and attached data loggers were housed in small Tupperware containers. Finally, canopy interception of precipitation was measured by attaching powder funnels to Nalgene bottles. Precipitation collectors were placed midway from the trunk to the dripline under focal study trees. Using these three microclimatic measurements, we found that scale herbivores altered both soil moisture and temperature and both herbivores altered crown interception of precipitation (Classen et al. 2005).

In conclusion, in a pinyon-juniper woodland, two herbivores significantly alter tree structure and physiology. These herbivore-driven changes at the tree level can result in important alterations to nutrient inputs and soil microclimates, yielding changed ecosystem processes and soil development. Using a combination of long-term herbivore-removal experiments and tree-tailored measurements, we were able to capture herbivore impacts, which may have been diffuse at the landscape scale.

SUMMARY

Forest canopies interact with the forest floor in a number of important ways. First, shading by the canopy reduces forest floor temperature and evaporation of soil water. Second, water moves upward from roots to the canopy via translocation within the plant and downward from the canopy via percolation of precipitation through the canopy. As water percolates through the canopy, it acquires additional material, so that water dripping from the canopy (throughfall) or flowing down boles (stemflow) show enhanced nutrient concentrations relative to raw precipitation. The reduced volume of water and protracted dripping from lower heights substantially reduce soil displacement and erosion, compared to soil displacement and erosion under open canopies. These transfers can be measured by using collectors and comparing volumes and nutrient concentrations and by means of sap-flux monitors and stable isotopes, each of which requires assumptions. Infrared gas analysis and eddy-covariance methods are becoming widely used to estimate water uptake via measurement of transpiration flux.

Material also moves from the canopy to the forest floor via translocation of photosynthates to roots and the rhizosphere and as litterfall. Litterfall consists of various plant and animal materials that vary in their contributions to forest floor biomass and nutrient flux. Translocation is measured most easily using stable isotopes, whereas litterfall can be measured using collectors on the forest floor. Litter components, such as whole leaves, fruits/flowers, wood, and animal feces, can be separated and analyzed separately for nutrients in order to assess their relative contributions to nutrient fluxes. Dead plant and animal material decomposes at rates determined by litter quality and by forest floor temperature and moisture. Decomposition and nutrient release can be measured using CO_2-absorption methods, stable isotopes, or mass and nutrient loss of weighed litter samples, all of which have advantages and limitations.

As discussed in previous chapters, the manipulation of forest canopies for rigorous experimental testing of hypotheses is difficult in mature forests, but several manipulative methods have been used. Long-term pretreatment measurement of unreplicated treatment plots can establish sufficient similarity to permit statistical comparison of posttreatment changes. Soil warming, using grids of heated cable, can simulate some effects of canopy opening. FACE experiments provide the opportunity to test the effects of increased atmospheric CO_2 on exchanges of material between the canopy and forest floor. Canopy trimming or cutting individual or groups of trees in replicated plots permits the evaluation of effects of canopy opening on canopy–forest floor interactions. Finally, manipulation of herbivory permits evaluation of herbivore effects on interactions between canopy and forest floor.

SUGGESTED READING

Classen, A. T., S. C. Hart, T. G. Whitham, N. S. Cobb, and G. W. Koch. 2005. Insect infestations linked to changes in microclimate: Important climate change implications. *Soil Science Society of America Journal* 69: 2049–57.

Fonte, S. J., and T. D. Schowalter. 2005. The influence of a neotropical herbivore (*Lamponius portoricensis*) on nutrient cycling and soil processes. *Oecologia* 146: 423–31.

Frost, C. J., and M. D. Hunter. 2007. Recycling of nitrogen in herbivore feces: Plant recovery, herbivore assimilation, soil retention, and leaching losses. *Oecologia* 151: 42–53.

Irvine, J., B. E. Law, P. M. Antoni, and F. C. Meinzer. 2002. Water limitations to carbon exchange in old-growth and young ponderosa pine stands. *Tree Physiology* 22: 189–96.

Nansen, C., T. Macedo, R. Swanson, and D. K. Weaver. 2009. Use of spatial structure analysis hyperspectral data cubes for detection of insect-induced stress in wheat plants. *International Journal of Remote Sensing* 30: 2447–64.

Shiels, A. B., J. K. Zimmerman, D. C. García-Montiel, I. Jonckheere, J. Holm, D. Horton, and N. Brokaw. 2010. Plant responses to simulated hurricane impacts in a subtropical wet forest, Puerto Rico. *Journal of Ecology* 98: 659–73.

8

TREETOPS AT RISK?

Engaging the Canopy Toolkit in Forest Conservation

INTRODUCTION: CANOPY SCIENCE AS A DRIVER FOR FOREST CONSERVATION

Despite extensive scientific research invested in tropical forests and their canopies during the past few decades, deforestation is increasing and rates of degradation continue to accelerate worldwide (Curran et al. 2004; Laurance and Perez 2006; Lowman 2009b). In some cases, the methods to measure forest degradation are not adequate to accurately quantify changes in canopy cover, especially with selective cutting or replacement of native stands with tree plantations (e.g., Puyravaud et al. 2010). Other studies have advocated efforts to save some tropical rain forests, but "we do not have to try to save every bit of it" (Janzen 2010). While anthropogenic drivers are increasingly linked to forest degradation, they can also lead to forest restoration. Forest canopy research, with its arsenal of creative methods and approaches to a challenging research environment, is becoming an important driver for conservation solutions. By applying not only the findings from canopy research but also the toolkit, canopy biologists are applying their expertise to critical issues regarding forest conservation.

CASE STUDIES OF APPLYING THE CANOPY TOOLKIT TO CONSERVATION

In this chapter, several case studies are described to illustrate new approaches to forest conservation utilizing canopy science: (1) quantifying forest ecosystem services as a driver for forest conservation, (2) engaging diverse stakeholders as stewards of forests,

(3) using canopy access for ecotourism and local income generation, (4) harvesting sustainable canopy products, and (5) using canopy science to inspire science education.

QUANTIFYING ECOSYSTEM SERVICES FROM FOREST CANOPIES

Several aspects of ecosystem services (reviewed in White et al. 2010) have emerged that prioritize canopy conservation. As rain forests continue to decline, the urgency of surveying their biodiversity becomes imperative. Reputedly, many small organisms dwelling in the treetops such as orchids as well as their resident invertebrates and other neighboring epiphytes, may have gone undetected simply because our ability to survey tree crowns has been limited to date. The wealth of biodiversity in the forest canopy is estimated at nearly half of the species on the planet (Wilson 1992), with no clear sense of absolute numbers, since canopy research is relatively new. Over the past 150 years, biologists have gradually expanded their knowledge of biodiversity as plant exploration became more extensive. The original estimates of 800,000 species made by Charles Darwin have been replaced by speculations ranging upward to 100 million (Wilson 1992); although estimates of perhaps 10–30 million are more frequently cited (Table 3.1). Our increasing knowledge of biodiversity on Earth has in great part been a consequence of tropical forest canopy exploration. Not only is the canopy important for species diversity; the types of species living in tropical treetops likely include important medicinal plants as well as genetic stock that may provide new sources of food, fabric, or other economic products.

BOX 8.1: THE ROLE OF PARAECOLOGISTS IN TWENTY-FIRST-CENTURY TROPICAL FOREST RESEARCH

Vojtech Novotny, George D. Weiblen, Scott E. Miller, and Yves Basset

Tropical forests, comprising numerous species interacting in complex ways, are just as challenging to study as genomes, with their complex interactions among numerous genes. The progress in genomics has been substantial over the past 50 years, as our ability to read and process digital information, including DNA sequences, has grown exponentially. The same cannot be said about the methods available for the study of tropical ecosystems. For example, most entomologists studying tropical insects today employ almost the same field methods and numbers of assistants as Alfred Wallace 150 years ago (Wallace 1869). Accordingly, our knowledge of tropical forest ecology has not progressed according to Moore's law. We have not cataloged all the species in any tropical forest ecosystem or mapped their interactions, and tropical biodiversity estimates remain uncertain (Hamilton et al. 2010).

More substantial progress in tropical ecology can be achieved either by developing new methods of ecosystem inventory and analysis or by applying existing methods more intensely. Both approaches are essential. The key methodological improvements in tropical ecology are coming mostly from molecular biology, which is at last becoming useful in ecological studies. Large-scale DNA sequencing programs improve species description and recognition (Hebert et al. 2004; Miller 2007), while molecular methods are becoming increasingly useful in mapping species interactions, including predation and parasitism (Greenstone 2006; King et al. 2008).

Tropical forest studies are often limited by inadequate sampling and the failures to record many species and interactions (Novotny and Basset 2000). Increasing sampling effort by one to three orders of magnitude would improve the situation, as illustrated by advances in understanding tropical forest dynamics based on detailed inventories of large forest plots. The first plot, established in 1982 on Barro Colorado Island (Hubbell and Foster 1983), mapped 240,000 stems less than 1 centimeter in diameter in 50 hectares of tropical forest. This represented a three-hundred-fold increase in sampling effort compared to the then standard protocol of mapping stems fewer than 5 centimeters in diameter in 1-hectare plots. Large plots have become the standard for tropical forest research across a global network of plots coordinated by the Center for Tropical Forest Science (CTFS; http://www.ctfs.si.edu). Arguably, similar increase in sampling is needed to advance our understanding of animals, particularly insects, in tropical forest ecosystems. This could be achieved by increasing research budgets or, more realistically, outsourcing fieldwork from academic professionals to paraecologists.

Paraecologists are essentially technicians supporting ecological research. They can efficiently organize field work; use various sampling methods; sort, preserve, document, and database specimens; and prepare samples for molecular analysis (Basset et al. 2000; Basset et al. 2004; Janzen 2004). Paraecologists, and research technicians in general, are able to develop specialized knowledge of research protocols, local ecosystems, or particular taxa over many years. They work well in synergy with postgraduate students, who lack comparably long-term work experience but can give paraecologists a broader and a more conceptual perspective on research.

In our experience, mostly in Papua New Guinea (PNG), the best paraecologists recruit either (1) from forest-dwelling communities, where they spent their formative years in the forest and acquired extensive traditional knowledge of plants and animals but typically had limited access to formal education, or (2) among secondary school or university graduates with interests in biology and the potential to become professional biologists. Some paraecologists move on to pursue graduate education and research careers and others seek long-term employment in their technical role.

Paraecologists are particularly efficient in large-scale or long-term sampling programs,

(continued)

(continued)

BOX 8.1, FIGURE 1. Parataxonomists created a "money tree" with their earnings from work on New Guinea's insect fauna.

in remote field conditions, and for sorting and processing large numbers of samples and specimens. For instance, eight paraecologists in PNG staged eight expeditions to remote rain forests over three years, during which they sampled and reared 75,000 caterpillars from 370 species. Their field work was sufficiently important to merit coauthorship of the resulting paper (Novotny et al. 2007). Likewise, our team of 18 paraecologists sampled 200,000 individual herbivores from 1,500 species and 11 guilds, feeding on more than 200 species of plants. This study documented almost 7,000 trophic interactions between particular plant and herbivore species. Although it required more than 50 person-years of work by paraecologists and additional 50 person-years by locally recruited field assistants, the study still suffered from insufficient sample size, as it documented less than 20 percent of the local plant–herbivore food web in the studied rain forest (Novotny et al. 2010).

The concept of paraecology was initially developed by Dan Janzen (Janzen 1992; Janzen 2004). In addition to Janzen's team in Costa Rica and ours in PNG, paraecologists

contributed, for instance, to surveys on the arthropods of La Selva (http://viceroy
.eeb.uconn.edu/ALAS/ALAS.html) and diversitas in the Western Pacific and Asia
(Darnaedi and Noerdjito 2008). They are employed at INBio (the National Biodiversity
Institute of Costa Rica; http://www.inbio.ac.cr/en/default.html) and throughout the
CTFS 50-hectare plot network, including our plot in PNG. Paraecologists make key
contributions to some of the largest data sets available for tropical ecosystems.

In spite of their success, few research projects involve paraecologists in a significant
way. Whereas many business activities are increasingly outsourced from developed
countries to developing countries where labor is cheap, ecological research continues to
import a work force of students and researchers from developed countries to locations
where fieldwork could be conducted more efficiently by paraecologists with local
knowledge of biological, political, and social circumstances. Significant and ongoing
investment of time and resources in paraecologist training is required to increase
efficiency, however.

The investment required means that paraecologist teams become cost-effective only
above a certain size and over longer periods of time. Five paraecologists working over
five years might represent a minimum threshold; although this of course depends on
the particulars of the project. For instance, Janzen's team in Costa Rica includes thirty
paraecologists and has been active for thirty years, whereas our team in PNG includes
twenty paraecologists and has been active for fifteen years. Few research institutions or
projects are prepared to make such commitment. A team of paraecologists has much
in common with other large-scale research programs: it is difficult to set up, expensive
to maintain, and takes a long time to pay off, but it significantly increases productivity.
Paraecologists have an important role to play in supporting the kind of large-scale and
long-term research that is needed to understand ecosystems as complex as tropical forests.

Canopy processes provide additional ecosystem services that are becoming recognized
at a global scale. The storage of carbon by trees, as part of their productivity, has been measured and valued in tropical forest ecosystems. REDD (Reduction of Emissions from Deforestation and Degradation) has achieved global recognition in climate change talks and
by the Intergovernmental Panel on Climate Change (IPCC), as warming temperatures
threaten the lives of millions of people and also threaten the integrity of many ecosystems
(reviewed in Lovejoy and Hanna 2005). Epiphytes and their inhabitants are integral to the
maintenance of rain forest ecosystems and have sometimes been referred to as the "canary
in the coalmines" for tropical forest health (Benzing 1990). The processes of nutrient cycling, energy production, water filtration, and many other essential functions of rain forests
are of global importance to the maintenance of life on our planet. Canopy processes drive
many of the ecosystem services provided by forests—water conservation, climate control,
productivity, carbon storage, and other invaluable benefits that forests provide to humans.

BOX 8.2: CANOPY RESEARCH AND FOREST POLICY

Andrew Mitchell

Why should canopy scientists care about forest policy? The answer is because without it there may be no forest canopy left for you to explore! Forests fall because they are worth more dead than alive in global markets. That means expanding agribusiness in Amazonia and Asia is the fastest growing cause of tropical deforestation globally. In Africa, and elsewhere, population expansion has left the poor with little option but to cut forests for food and firewood. To combat these causes of deforestation, forests need to be worth more standing up than cut down. We are at an extraordinary moment in human history when that might just be about to occur. More than this, it is canopy scientists who have played a key role in making it happen.

The forest canopy is where life meets the atmosphere. Canopy researchers over the last thirty years have uncovered not only that almost half of all life lives up there but that this biodiversity provides ecosystem services to humanity on an epic scale. Instruments on canopy towers across the Amazon have revealed how forests help regulate the carbon and water cycle of our planet. We now know that tropical forests absorb about 1 tonne of carbon per hectare per year, an atmospheric pollution cleaning service worth billions of dollars per year. And they do it for free! Conversely, when forests are cleared and burned for land, we reduce this service, and they release a giant store of carbon, annually emitting more to the atmosphere than the entire global transport sector.

This knowledge has provided the inspiration for Reducing Emissions from Deforestation and Degradation (REDD) in developing countries. It's a proposed mechanism under the UN Framework Convention on Climate Change (UNFCCC), which would pay countries for combating deforestation through selling carbon credits to local or international emitters. It is complicated because in many countries, who owns the forest, and therefore who should receive the funds, is not clear, and in some countries institutions are weak and officials are poorly paid, so they may accept payments to falsify papers. If REDD can be made workable, it offers for the first time a billion-dollar reason for governments to keep their forests standing instead of cutting them down for profit.

But paying for the carbon service is just the start. The forest canopy of Amazonia recycles 8 trillion tonnes of water into the atmosphere every year. Canopy scientists have tracked this moisture and have found that it is recycled vast distances across the forest to the Andes. where some falls as snow, and also speculate that it carries on south to feed the agricultural heartlands of southern Brazil and possibly as far south as the breadbasket of the La Plata Basin. The Congo Basin's forests may prove equally important to rainfall distribution in Africa.

Scientists lobbied the UN to make sure that REDD recognizes such services, too, as well as the value the canopy provides for global biodiversity and the provision of livelihoods for forest peoples as well. So REDD is now called REDD+, where the plus embraces these other services, beyond the carbon cycle alone.

Forests and their canopies are in fact a kind of proxy for the way in which nature underpins wealth creation globally. The services they provide underpin climate, food, energy, health, and livelihood security for a poor forest family to a major agribusiness or food investor. A bug may not be worth a billion dollars, but the services of the entire system that the bug is a part of could be; conveying that to policy makers has caused them to sit up and listen.

For REDD to work, a government must prove it is slowing emissions from deforestation. Satellites will monitor fires and the forest area regularly. The UN will authorize payments based on performance. Measures must be taken to prevent deforestation "leaking" to another state or another country. New regulations will need to be agreed upon to make forest carbon credits part of the international carbon market, so that polluters can buy them

(continued)

BOX 8.2, FIGURE 1. Andrew Mitchell and canopy scientists with Prof. Virgilio Viana, state secretary of the environment, and staff, 60 meters up a canopy tower in the Amazon, launching the Forest NOW declaration in 2007. The declaration, signed by 500 signatories, including heads of state and Nobel Prize–winning scientists, was presented at the UN Bali conference on climate change and helped persuade policy makers to include forests in the climate talks for the first time, leading to REDD+ today.

(continued)

to offset emissions they cannot avoid releasing themselves. This could generate billions of dollars annually for forest-owning nations. Safeguards must be put in place to ensure that communities that own forests, as well as governments, share equitably in the funds raised. Nations with lots of ancient forests who are not burning them need also to be given incentives to maintain these forests. Others that are restoring their forests may get rewarded for doing so, because of the carbon they will take out of the atmosphere as they grow. To find out more, visit http://www.theREDDdesk.org.

ENGAGING DIVERSE STAKEHOLDERS AS STEWARDS OF FORESTS

The conventional stakeholders of tropical forests are big government and conservation nongovernmental organizations (NGOs), but the fact that tropical forests continue to decline on a global scale indicates that these stakeholders are not always effective. Throughout many tropical regions, however, other stakeholders exist who have successfully conserved local forests. Increased communication of their success stories could serve as models for effective forest conservation solutions and subsequently reverse the current declines in forest resources.

One emerging group of forest conservation stakeholders is farmers. In Africa, "evergreen agriculture" combines the principles of conservation farming with the establishment of "fertilizer trees" as part of a new agroforestry movement (Garrity 2010). The canopy sheds its nutrient-rich leaves and subsequently fertilizes the otherwise poor soils for crops beneath. Such new, innovative practices could not only dramatically improve the quality of life for millions of people but also conserve local forests and native canopy. Eighteen countries in Africa are currently engaged in trials that involve evergreen agriculture. Australia similarly experienced widespread forest diebacks that led to the death of livestock and subsequent economic declines, thereby inspiring graziers to restore stands of native forest (Lowman and Heatwole 1992).

Religious leaders comprise another successful group of stakeholders, whose efforts to conserve forests are exemplary at a global scale (Bossart et al. 2006). One notable case study is the Coptic or Christian Orthodox church in Ethiopia (Jarzen et al. 2010; Lowman 2011; Wassie-Eshete 2007). Approximately 35,000 churches ranging from 3 to 1,500 hectares exist throughout the country, dating back to 360 AD, and each is surrounded by a tract of native forest, since biodiversity stewardship is part of the church mission (Fig 8.1A). These forest patches not only preserve biodiversity but also provide numerous ecosystem services: pollinators, native seed stock, shade, spiritual sites, medicines from the plants, and freshwater conservation through both rainfall patterns and the existence of underground springs (Jarzen et al. 2010; Lowman 2011).

Despite the church's authority as an admired and strong national leader, pressure from subsistence agriculture and also from the simple demands for firewood threaten these tiny green jewels dotting Ethiopia's otherwise brown, arid landscape (Fig 8.1A). With funding from National Geographic and Tree Research, Education, and Exploration (TREE) Foundation, a group of conservation biologists are working with the clergy to empower their leadership in educating the locals about ecosystem services. With a focus on insect pollinators as indicator species of forest health and utility, workshops were hosted with the local clergy to educate them about the natural capital of their church forests (Lowman 2011; Fig 8.1B). With the enthusiastic support of the clergy, environmental education programs are under way for local youth to sample insects and thus understand the importance of forest conservation in their regions. Armed with insect nets, vials, and climbing ropes, the pilot programs attracted crowds of kids in each church forest, and they marveled at the six-legged creatures swept from their foliage (Figure 8.1C). Despite language barriers, the church provided a common bond for working together to conserve part of the natural heritage of Ethiopia. Many of these canopy insects are likely to be labeled new species (Fig. 8.1D). Loss of these last remaining patches of forest would represent extinction for many native trees, insects, birds, and mammals that use these church forests as refugia. Ethiopia has lost more than 95 percent of its forest cover, but the partnership of religion and science has the capacity to save the remaining 5 percent and perhaps ultimately lead to restoration practices.

USING CANOPY ACCESS TO BOOST ECOTOURISM AND LOCAL ECONOMIES

Canopy access techniques have proven exemplary in providing ecotourism in tropical regions, thereby boosting the economy of local indigenous villages through ecotourism (reviewed in Cardelús et al. 2012). In many tropical regions, the payments derived from logging operations far exceed any economic benefits from conservation; in fact, a "conservation royalty scheme" has been suggested to provide business incentives to retain forests (Novotny 2010). Canopy access tools—namely, walkways—create a mechanism to examine the socioeconomic plus scientific metrics of forest conservation through sustainable economy, specifically ecotourism. Business ventures involving canopy exploration include bird-watching, education-based nature tours, spas, and holistic medicine (Weaver 2001). These "green businesses," when integrated with basic research on forest canopies, provide sustainable economies for local villagers (Lowman 2009a).

Canopy walkways range in cost from $100 to $3,000 per meter and generate revenues for local stakeholders; they also provide environmental education to a broad visitorship (Lowman 2004b; Lowman and Bouricius 1995). More than twenty canopy walkways currently operate in tropical forests around the world, serving research, education, and ecotourism (Lowman 2009b). In the Sucasari tributary of the Rio Napo in western Peru, the world's longest canopy walkway provides employment for more than a hundred local

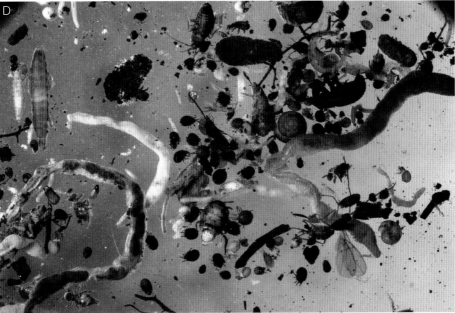

FIGURE 8.1.
Church forests in Ethiopia, where religious leaders are the stakeholders for canopy conservation and preservation of forest ecosystem services, including: (A) an aerial landscape of church forests dotting the agricultural landscape of northeastern Ethiopia; (B) a workshop of Coptic church leaders to discuss church forest conservation, with the church taking on an important leadership role in preserving forest canopies; (C) children from local villages learn to monitor Ethiopian church forest insect biodiversity, which includes resources of honey and pollination; and (D) collections representing new species of arthropods from Ethiopian forest canopies.

families and helps educate thousands of western visitors every year about rain forest ecology and conservation (Lowman 2009b). Costing some $250,000 to build, it generates revenue estimated at $1.2 million per year and most important, provides an economic incentive to conserve the primary forest.

In Western Samoa, a canopy access platform was similarly constructed, enabling local villagers on the Island of Savaii'i to pay for their new school via ecotourism profits, instead of profits from selling logs. Led by ethnobotanist Paul Cox, a team of American canopy builders (http://www.treefoundation.org) met with the fifteen village chiefs to discuss the risks of this ecotourism venture. After drinking copious amounts of kava, the chiefs unanimously approved the construction of a walkway into an emergent Ficus tree. Profits from this successful venture paid for school construction without the more conventional "solution" of logging local forests (Lowman et al. 2006). Over a short two-year period, the villagers paid back a $53,000 loan for their school construction by leading canopy tours. The chiefs wisely recognized that logging would ultimately destroy their island ecosystems and that ecotourism would provide a more sustainable income stream.

PROMOTING FOREST CONSERVATION WITH SUSTAINABLE CANOPY PRODUCTS

Sustainable products from rain forest canopies range from iguana meat to tropical butterflies for temperate museum exhibits to orchid farming. In Cameroon, Africa (Lowman et al. 2006), local pygmy villagers were trained to farm epiphytes in the canopy instead of sell logs to foreign logging teams (Lowman 2009a). Although this new venture is in its infant stages, the notion of creating sustainable products from forests without harvesting their timber is critical. In Belize, Ian Meerman directs a butterfly farming operation, whereby young women from local villages are trained to rear and care for caterpillars of tropical butterflies, which are sold to international museums for butterfly gardens. This business not only encourages sustainable practices for tropical butterflies but also empowers women with paychecks and ultimately lowers the birth rate.

Almost every tropical port, town, or hotel offers crafts that are products of the local forests—shade-grown coffee, wooden trinkets, rattan furniture, or palm baskets. Such efforts utilize relatively small amounts of forest products but have the capacity to command relatively high prices as collectibles. A review of forest-sustainable products is forthcoming (Schowalter 2012).

USING CANOPY RESEARCH TO INSPIRE SCIENCE EDUCATION

Canopy Conferences as a Platform for Education Outreach

For the first time ever, the fifth International Canopy Conference (ICC) in 2009 was convened in an emerging country, hosted by Ashoka Trust for Research on Ecology and the Environment (ATREE) in Bangalore, India. Also, for the first time ever, education outreach was added as a major component of the conference agenda, including a highly popular canopy education workshop that drew Indian students, teachers, scientists, and stakeholders from all regions of the country. This session was advertised as an open forum where laypersons and teachers could interface with canopy scientists, who enthusiastically shared their experiences, taking science outside the conventional walls of academia. In a country like India, this was groundbreaking. One year later, the outreach programs inspired by this session illustrate the growing importance of canopy science as a platform for science education, in India as well as elsewhere.

The education outreach sessions at ICC fostered subsequent outreach programs by canopy scientists throughout India, including (1) tree-planting programs in schoolyards, (2) citizen surveys of butterflies, (3) Fulbright funding for canopy education outreach, (4) participation by canopy scientists in Earthwatch field courses for HSBC Bank employees, (5) graduate training programs in canopy access, (6) publication of a canopy science book for the public, and (7) a certificate course in conservation for young professionals. All these activities illustrate the genesis of a new culture among Indian canopy scientists to engage in education outreach. Only one year later, hundreds of Indian students and thousands of citizens have benefited directly from the canopy conference's education session (Lowman, Devy, and Ganesh, in press).

The emerging priorities of communicating science and fostering education of diverse stakeholders is an increasingly important platform for scientific conferences, especially given the urgent priorities for STEM (science, technology, engineering, and mathematics) initiatives in many countries. In India, the priority of education outreach at the canopy conference provided a new model for scientific meetings. This session fostered discussion of a critical question: Will canopy scientists pursue business as usual in the face of significant forest degradation or challenge themselves to do things differently, such as prioritizing the education of youth, policy makers, and citizens? Keynote speaker Dr. Thomas Lovejoy (former President, Heinz Center, Washington, DC) remarked to the international attendees, "By any measure, tropical forests are in big trouble." Will education outreach, as an innovative approach to engage diverse stakeholders, inspire metrics of success for forest canopy conservation?

In India, no conference session attracted such an abundance and diversity of attendees as education outreach. More than one hundred teachers of India's southern states representing diverse cultures, ages, genders, and religions eagerly listened to important take-home messages: (1) develop short courses in canopy science for teachers, citizens, and K–12; (2) create climate change awareness in schools and at the level of regional government; (3) share resources; and (4) plant trees to "green India" as a strategy to offset climate change and sustainable services.

Historically, India has seen economics (not environment) dominate its policy-making decisions, to the detriment of some of its natural resources. The ICC fostered a dialogue where a combination of economics and environment, via more effective science communication and education outreach, may hold the key to producing a scientifically literate generation of citizens and policy makers. If every practicing scientist were to tithe 10 percent of his or her professional time to education outreach, not exclusively limited to technical publications, then perhaps the notion of a scientifically literate public could be attained. Currently, in India and elsewhere, many science professionals still think that education outreach does not fall into their purview. The expanding footprint of education outreach sessions at scientific conferences provides one metric of hope for improving science literacy among diverse stakeholders, not just scientists.

Virtual Technologies to Advance STEM Education

One example of the application of canopy research to science education is the Jason Expedition (http://www.jason.org), a distance learning program that annually engages more than 3 million middle school students and teachers worldwide on remote field expeditions using satellite technology (Lowman 1999). In 1994, 1999, and 2004, the Jason Project featured canopy research, including the challenges of access and data collection in the treetops. During 2004, Jason XV (expedition 15, see http://www.jason.org) studied the links between the brown (i.e., forest floor) and green (i.e., canopy) food webs in the forest canopy of the Smithsonian Institute's tropical research site called Barro Colorado Island, Panama (Lowman et al. 2006). Science curricula developed specifically for the canopies of Panama fostered student learning about the complex linkages among biodiversity, biogeochemical cycling, and global environmental conditions. Researchers conducted canopy studies that were broadcast live into classrooms throughout the world, providing a unique model that integrates research with ecology education. Middle school students participated in data collection that led to scientific publications coauthored with middle school youth (e.g., Burgess et al. 2003). Additional educational programs include National Geographic television, Discovery Channel, numerous museums with informal science education exhibits (e.g., the American Museum of Natural History's biodiversity exhibit and the North Carolina Museum of Natural Science's Nature Research Center), and the plethora of books and websites now available for youth and school groups. Special ecotourism expeditions are aimed at teenagers, and numerous educational curricula include applications for mobile handheld devices, MP3s of tropical bird songs, ringtones of tree frogs, and other ways that educate users in a recreational fashion (Fig. 8.2).

The next generation of canopy researchers is growing up in a world of sound bites, virtual nature, and emerging technologies that can "transport" them to different ecosystems from a handheld device. While many educators currently worry about what is termed the "nature-deficit disorder" (Louv 2005; Lowman and Mourad 2010), the opportunities to blend technology and in situ natural systems will hopefully inspire these new forest scientists to find sound solutions to global environmental challenges. Students are not only capable of quantifying biodiversity in forest canopies, but they can model the impacts of climate change or outbreaks to better understand how to manage the finite resources of our forests.

FIGURE 8.2.

The new Nature Research Center at North Carolina's Museum of Natural Sciences integrates the "virtual" and "real" nature of forest canopies and other cutting-edge research to inspire the next generation of scientists.

Education about ecosystem services has inspired a new generation of rain forest conservationists—youth who work on behalf of REDD and climate change (http://www .youngvoicesforclimatechange.org), young people advocating for conservation of lemurs and three-toed sloths (http://www.treefoundation.org), school and church groups who sell shade-grown coffee, and consumers who apply strict guidelines for certified timber or sustainably grown produce. But until the power of consumers becomes absolutely and directly linked to rain forest exploitation, the degradation of forest canopies is likely to continue. And every additional increment of forest destruction and fragmentation magnifies into a more significant and irreversible situation. It is still very much a chain-saw versus canopy walkway situation, with the consumer in control. The continued escalation of diverse stakeholders for forest conservation, in addition to the efforts of big government and NGOs whose track records in the late twentieth century were not effective enough to reverse major degradation, may be key to achieving success in reversing deforestation trends.

SUMMARY

After forty years of serious development, canopy science is still an emerging frontier of exploration promising not only adventure and scientific discovery but also solutions to global challenges. Important environmental issues such as climate change, biodiversity conservation, and forest ecosystem services have inspired data collection within canopies as well as above and below canopies and catalyzed large-scale ecological monitoring. But perhaps most important, canopy ecology may herald a new set of achievement metrics for the next generation of ecologists. As canopy science matures, the treetops have become scientific, economic, and social drivers for outreach in education and conservation. Rather than relying on technical scientific publications and data sets as the only measures for promotion and tenure, the next generation of ecologists increasingly prioritizes education outreach and applied conservation as metrics indicative of success as a scientific professional. The integration of research, education outreach, and conservation as the new toolkit for young canopy ecologists may lead to better stewardship of forests.

9

CONCLUSIONS AND RECOMMENDATIONS

Canopy science has advanced greatly from its early days of individual researchers climbing into accessible trees using ropes or ladders and collecting what could be reached or using binoculars to observe more distant objects. Today we have a variety of canopy access methods and sophisticated measuring devices that can record continuously by means of solar battery power. Canopies can now be observed remotely, at a global scale, using satellite imagery. However, there are a number of challenges remaining:

1. Scaling of data from individual leaves to regional canopies, necessary to validate and correlate remotely sensed fluxes with more directly measured data at the leaf scale

2. Canopy access platforms representing more forest types and regions

3. Standardization of methods to facilitate comparison of data across regional and global canopies

4. Studies of effects of vertical and horizontal variation in canopy depth and structure

5. Experimental manipulation of canopy structure and processes to test hypotheses

6. Public education to build support for conservation of canopy functions

Advances in these areas will greatly improve our understanding of, and incentive to conserve, forest canopies and their roles in global processes that control temperature, precipitation, and air quality that, in turn, affect ecosystem services on which we depend.

SCALING OF DATA FROM LEAF TO GLOBAL CANOPY

We recommend continued efforts to correlate and validate relationships among data from individual leaf to global canopy scales in order to improve confidence in data from remote-sensing studies. Scaling of ecological data from small scales at which data can be collected most accurately and conveniently to larger scales has been an issue for decades. Forest ecologists wrestled with the problem of calculating whole-tree and whole-canopy photosynthesis and respiration from measurements at the individual leaf scale during International Biological Programme (IBP) days (Gutschick 1996).

In the earliest attempt to correlate multiscale data, Odum and Jordan (1970) enclosed a multitree patch of tropical rain forest in a plastic bag, with provision for necessary ventilation, and measured whole-canopy photosynthesis, respiration, and net primary production, comparing their results with measurements from individual leaves. More recently, modeling approaches have proven useful for testing and validating correlations between leaf- and canopy-scale measurements (Furon et al. 2007; Ogawa 2008; Xu and Griffin 2008).

CANOPY ACCESS PLATFORMS

We recommend additional canopy platforms (towers, cranes, walkways, or other permanent access facilities) to represent a greater variety of forests, especially forest types and regions that currently are underrepresented. This also will require commitment for maintenance by sponsoring agencies. The Wind River Canopy Crane Research Facility in Washington State recently was disassembled due to budget cuts. This had been a premier facility in one of the most productive and pristine old-growth (five hundred years old) coniferous forests in North America. Its loss leaves North America with only one crane operated by the Smithsonian Institution in an eastern deciduous, second-growth forest. Although expensive, cranes permit the expansion of research beyond the footprint of the tower, as described in Chapter 3.

Regions that would benefit most from additional canopy platforms to facilitate access for research include major portions of the tropics, in Africa, Asia, and South America. Temperate forests in South America also are underrepresented. In addition, the extensive boreal forests of Siberia and Canada have relatively few platforms for canopy access.

STANDARDIZATION OF METHODS

We recommend a standardization of methods to facilitate comparison of forest structures, diversity, and processes among forest types and regions for testing hypotheses concerning controlling factors and effects on other environmental conditions. We have not attempted to recommend particular methods but rather synthesized available methods and evaluated their advantages and disadvantages for various objectives. For example, interception traps provide data on the diversity and relative abundances of flying insects, integrated over the time between trap collection, but provide no data on density relative to tree resources that would be necessary to calculate food resources available to gleaning birds or herbivore effect on nutrient fluxes (see Chapter 5). Attaining the density data necessary for calculating food availability or herbivore effects on nutrient fluxes requires methods such as branch bagging that provide numbers of insects or herbivores per habitat unit (Chapter 5).

We encourage the collaboration of canopy researchers across disciplines and regions, to select and use appropriate standardized methods to meet particular objectives. Examples of such collaborations include the International Canopy Network (ICAN), International Long Term Ecological Research (ILTER) Network, and the AmeriFlux and EuroFlux networks. These teams collect data using standardized methods that facilitate comparison among forest types and regions. Such data permit the testing of hypotheses that could not be tested with noncomparable data resulting from use of dissimilar methods.

EFFECTS OF VERTICAL AND HORIZONTAL VARIATION

We recommend further study of the effects of vertical and horizontal variation in canopy depth and structure on gradients in temperature and moisture profiles, species abundance, carbon and water fluxes, and regional climate. Deeper and denser canopies apparently affect these gradients to a greater extent than do shorter or sparser canopies, based on current data (Foley et al. 2003). How do various forest types affect these gradients? How do trees allocate water among various branches, and how does this affect relative humidity gradients within the canopy? Water flux has been measured only on the main trunk of individual trees. Presumably, water flux within the crown depends on the rate of evapotranspiration, with high flux rates providing adequate water and nutrients for continued growth and photosynthesis and low flux rates limiting growth and photosynthesis and leading to branch death. How do various organisms respond to gradients in microclimate and habitat structure, and what are the consequences for canopy processes, such as primary productivity and herbivory? How does variation created by environmental gradients and disturbance across time and space affect canopy capacity to modify these gradients in different forest types? More particularly, how does anthropogenic modification of canopy structure through deforestation or plantation forestry affect canopy functions and capacity to modify regional and global climate (Foley et al. 2003; Janssen et al. 2008)?

EXPERIMENTAL MANIPULATION

We recommend application of experimental manipulations at all scales of canopy study in order to improve our understanding of factors controlling forest canopies and effects of canopy changes on environmental conditions. As noted throughout this book, experimental manipulation is necessary to test hypotheses concerning controlling factors and effects of canopy changes unambiguously. Forests represent a particular challenge in this regard because of their complex three-dimensional structure and the labor costs and safety hazards involved in manipulation. Single paired trees or plots provide insufficient data for statistical comparison of differences (Hurlbert 1984). In some cases, where comparable methods and data are available, multiple paired tree or plot data can be combined in a meta-analysis to test hypotheses (e.g., Jactel and Brockerhoff 2007; Vehvilainen et al. 2007). If sufficient pretreatment data are available to demonstrate substantial similarity among plots prior to treatment (Hurlbert 1984), departure from similarity following a treatment or environmental change can be evaluated (e.g., Davidson et al. 2008; Hurlbert 1984; Nepstad et al. 2007). However, manipulations at the branch, whole-crown, and whole-canopy levels have demonstrated the value of rigorously designed experiments.

Manipulation at the branch level has been used to test hypotheses concerning foliage and branch density, branch orientation, and other aspects of crown architecture. Examples include experimental manipulation of branch spacing and foliage density within tree crowns to evaluate effects of these factors on bird and spider activity (De Souza and Martins 2005; Whelan 2001); experimental manipulation of the height of floral resources from different canopy strata to demonstrate that the apparent fidelity of pollinator species to particular canopy strata reflected pollinator preferences for particular floral resources (Roubik 1993); experimental placement of artificial soil habitats at various heights within the canopy to assess the effect of soil volume, soil moisture, and canopy height on the development of oribatid assemblages (Lindo et al. 2008); and manipulation of predator and leaf litter abundance in natural and artificial tree holes to evaluate effects on community development (Yanoviak 2001).

More ambitious manipulations at the stand level have also been conducted. Examples include replicated plots thinned to designated stem retention levels in aggregated or dispersed patterns to evaluate the effects of canopy opening on forest structure and function (Schowalter et al. 2005) and replicated plots thinned to designated overstory or understory densities and burned or unburned in a factorial design to evaluate the effects of restoration treatments in mixed-conifer old-growth forests on multiple forest structure and ecosystem variables (North et al. 2002; North et al. 2007). In one particularly ambitious project, climbers cut all canopy branches smaller than 10 centimeters in diameter in replicated plots and redistributed the resulting cut debris among trimmed or untrimmed plots to simulate the effects of canopy opening and redistribution of canopy biomass to the forest floor by hurricanes on ecosystem variables (Richardson et al. 2010; Shiels et al. 2010).

PUBLIC EDUCATION

We recommend increased effort to involve the general public in canopy ecotourism and educational programs that will support conservation of forests. Data summarized in this book and others (e.g., Basset et al. 2003; Lowman and Rinker 2004) document the importance of forest canopies for the maintenance of biodiversity and global environmental conditions that support ecosystem services on which humanity depends. Clearly, further deforestation or harvest of primary forests and replacement by young, secondary (often planted) forests will alter global carbon flux, at the same time that fossil fuel combustion also is contributing to increased atmospheric CO_2 concentration, as well as albedo, airflow, and other factors that control global temperature, precipitation, and atmospheric and oceanic circulation patterns. These changes jeopardize the ability of forests to provide food, fiber, timber, fresh water, and other ecosystem services.

Public education is critical to building support for the conservation of forest canopies and the services they provide. Ecotourism in forest canopies will not be sufficient by itself to build support for conservation. Educational programs are necessary to emphasize the importance of forest canopies to continued quality of life. Additional canopy access platforms and educational programs in the canopy, such as the Jason Project, will enhance appreciation for canopy complexity and canopy contributions to global climate.

REFERENCES

Addor, E. E., W. N. Rushing, and W. E. Grabau. 1970. A procedure for describing the geometry of plants and plant assemblages. In *A tropical rain forest*, ed. H. T. Odum and R. F. Pigeon, 151–67. Oak Ridge, TN: US Atomic Energy Commission.

Agee, J. K. 1993. *Fire ecology of Pacific Northwest forests*. Washington, DC: Island Press.

Aldrich, R. C., and A. T. Drooz. 1967. Estimated Frazer fir mortality and balsam woolly aphid infestation trend using aerial color photography. *Forest Science* 13: 200–313.

Allee, W. C. 1926. Measurement of environmental factors in the tropical rain-forest of Panama. *Ecology* 7: 273–302.

Allen, C. D., and D. D. Breshears. 1998. Drought-induced shift of a forest-woodland ecotone: Rapid landscape response to climate variation. *Proceedings of the National Academy of Sciences USA* 95: 14839–42.

Amaranthus, M. P., and D. A. Perry. 1987. Effect of soil transfer on ectomycorrhiza formation and the survival and growth of conifer seedlings on old, nonreforested clear-cuts. *Canadian Journal of Forest Research* 17: 944–50.

Amiro, B. D., A. G. Barr, T. A. Black, R. Bracho, M. Brown, J. Chen, K. L. Clark, et al. 2010. Ecosystem carbon dioxide fluxes after disturbance in forests of North America. *Journal of Geological Research—Biogeosciences* 115, G00K02. doi:10.1029/2010JG001390.

Anderson, D. L. 2009. Ground vs. canopy methods for the study of birds in tropical forest canopies: Implications for ecology and conservation. *The Condor* 111: 226–37.

Anderson, J. E., L. C. Plourde, M. E. Martin, B. H. Braswell, M. L. Smith, R. O. Dubayah, M. A. Hofton, and J. B. Blair. 2008. Integrating waveform lidar with hyperspectral imagery for inventory of a northern temperate forest. *Remote Sensing of Environment* 112: 1856–70.

Angulo-Sandoval, P., H. Fernández-Marín, J. K. Zimmerman, and M. T. Aide. 2004. Changes in patterns of understory leaf phenology and herbivory following hurricane damage. *Biotropica* 36: 60–67.

Anzures-Dadda, A., and R. H. Manson. 2007. Patch- and landscape-scale effects on howler monkey distribution and abundance in rainforest fragments. *Animal Conservation* 10: 69–76.

Appanah, S. 1990. Plant-pollinator interactions in Malaysian rain forests. In *Reproductive Ecology of Tropical Forest Plants*, ed. K. Bawa and M. Hadley 85–100. Paris: UNESCO/Parthenon.

Appanah, S., and H. T. Chan. 1981. Thrips: Pollinators of some dipterocarps. *Malaysian Forester* 44: 234–52.

Arriaga-Weiss, S. L., S. Calmé, and C. Kampichler. 2008. Bird communities in rainforest fragments: Guild responses to habitat variables in Tabasco, Mexico. *Biodiversity Conservation* 17: 173–90.

Art, H. W. 1993. *The dictionary of ecology and environmental science*. New York: Holt.

Asner, G. P., R. F. Hughes, P. M. Vitousek, D. E. Knapp, T. Kennedy-Bowdoin, J. Boardman, R. E. Martin, M. Eastwood, and R. O. Green. 2008. Invasive plants transform the three-dimensional structure of rain forests. *Proceedings of the National Academy of Sciences of the United States of America* 105: 4519–23.

Asner, G. P., D. E. Knapp, T. Kennedy-Bowdoin, M. O. Jones, R. E. Martin, J. Boardman, and C. B. Field. 2007. Carnegie Airborne Observatory: In-flight fusion of hyperspectral imaging and waveform light detection and ranging (wLiDAR) for three-dimensional studies of ecosystems. *Journal of Applied Remote Sensing* 1: Article number 013536. doi:10.1117/1.2794018.

Asner, G. P., and R. E. Martin. 2011. Canopy phylogenetic, chemical and spectral assembly in a lowland Amazonian forest. *New Phytologist* 189: 999–1012.

Asner, G. P., and P. M. Vitousek. 2005. Remote analysis of biological invasion and biogeochemical change. *Proceedings of the National Academy of Sciences of the United States of America* 102: 4383–86.

Baker, L. R., and O. S. Olubode. 2007. Correlates with the distribution and abundance of endangered Sclater's monkeys (*Cercopithecus sclateri*) in southern Nigeria. *African Journal of Ecology* 46: 365–73.

Baldocchi, D. D. 2008. "Breathing" of the terrestrial biosphere: Lessons learned from a global network of carbon dioxide flux measurement systems. *Australian Journal of Botany* 56: 1–26.

Baldocchi, D. D., J. Finnigan, K. Wilson, K. T. Paw U, and E. Falge. 2000. On measuring net ecosystem carbon exchange over tall vegetation on complex terrain. *Boundary-Layer Meteorology* 96: 257–91.

Baldocchi, D. D., B. B. Hincks, and T. P. Meyers. 1988. Measuring biosphere-atmosphere exchanges of biologically related gases with micrometerological methods. *Ecology* 69: 1331–40.

Basset, Y. 1997. Species-abundance and body size relationships in insect herbivores associated with New Guinea forest trees, with particular reference to insect host specificity. In *Canopy arthropods*, ed. N. E. Stork, J. Adis, and R. K. Didham, 237–64. Boca Raton, FL: Chapman and Hall.

Basset, Y., B. Corbara, H. Barrios, P. Cuénoud, M. LePonce, H.-P. Aberlenc, J. Bail, et al. 2007. IBISCA-Panama, a large-scale study of arthropod beta-diversity and vertical stratification in a lowland rainforest: Rationale, study sites and field protocols. *Entomologie* 77: 39–69.

Basset, Y., O. Missa, A. Alonso, S. E. Miller, G. Curletti, M. de Meyer, C. Eardley, W. T. Lewis, M. W. Mansell, V. Novotny, and T. Wagner. 2008. Choice of metrics for studying arthropod responses to habitat disturbance: One example from Gabon. *Insect Conservation and Diversity* 1: 55–66.

Basset, Y., V. Navotny, S. Miller, and R. Kitching, eds. 2003. *Arthropods of tropical forests*. Cambridge: Cambridge University Press.

Basset, Y., V. Novotny, S. E. Miller, and R. Pyle. 2000. Quantifying biodiversity: Experience with parataxonomists and digital photography in Papua New Guinea and Guyana. *BioScience* 50: 899–908.

Basset, Y., V. Novotny, S. E. Miller, G. D. Weiblen, O. Missa, and A. J. A. Stewart. 2004. Conservation and biological monitoring of tropical forests: The role of parataxonomists. *Journal of Applied Ecology* 41: 163–74.

Bawa, K. S., M. D. Lowman, and N. M. Nadkarni. Forthcoming. Forest canopies as Earth's support systems: Priorities for research and conservation. *Biotropica*.

Beebe, W. 1949. *High jungle*. New York: Duell, Sloan, and Pearce.

Beehler, B. 1991. *A naturalist in New Guinea*. Austin: Texas University Press.

Benzing, D. 1990. *Vascular epiphytes*. Cambridge: Cambridge University Press.

Binkley, D. 1999. Disturbance in temperate forests. In *Ecosystems of the world 16: Ecosystems of disturbed ground*, ed. L. R. Walker, 453–66. Amsterdam: Elsevier.

Birnbaum, P. 2001. Canopy surface topography in a French Guiana forest and the folded forest theory. *Plant Ecology* 153: 293–300.

Blanton, C. M. 1990. Canopy arthropod sampling: A comparison of collapsible bag and fogging methods. *Journal of Agricultural Entomology* 7: 41–50.

Bohlman, S. A., and S. W. Pacala. 2012. A canopy layering model that demonstrates regulation of crown structure and partitions dynamic rates in a tropical forest. *Journal of Ecology* 100: 508–18.

Bongers, F. 2001. Methods to assess tropical rain forest canopy structure: An overview. *Plant Ecology* 153: 263–77.

Bossart, J. L., E. Opuni-Frimpong, S. Kuudaar, and E. Nkrumah. 2006. Richness, abundance, and complementarity of fruit-feeding butterfly species in relict sacred forests and forest reserves in Ghana. *Biodiversity and Conservation* 15: 333–59.

Bourguignon, T., M. Leponce, and Y. Roisin. 2009. Insights into the termite assemblage of a neotropical rainforest from the spatio-temporal distribution of flying alates. *Insect Conservation and Diversity* 2: 153–62.

Bouricius, W. G., P. K. Wittman, B. A. Bouricius. 2002. Designing canopy walkways: Engineering calculations for building canopy access systems with cable-supported bridges. *Selbyana* 23(1): 131–36.

Bowling, D. R., N. G. McDowell, J. M. Welker, B. J. Bond, B. E. Law, and J. R. Ehleringer. 2003. Oxygen isotope content of CO_2 in nocturnal ecosystem respiration: Observations in forests along a precipitation transect in Oregon, USA. *Global Biogeochemical Cycles* 17(31): 1–14.

Brauman, K. A., D. L. Freyberg, and G. C. Daily. 2010. Forest structure influences on rainfall partitioning and cloud interception: A comparison of native forest sites in Kona, Hawai'i. *Agricultural and Forest Meteorology* 150: 265–75.

Bray, J. R., and E. Gorham. 1964. Litter production in forests of the world. *Advances in Ecological Research* 2: 101–57.

Brenes-Arguedas, T., P. D. Coley, and T. A. Kursar. 2008. Divergence and diversity in the defensive ecology of *Inga* at two Neotropical sites. *Journal of Ecology* 96: 127–35.

Brokaw, N. V. L. 1982. The definition of a treefall gap and its effect on measures of forest dynamics. *Biotropica* 14: 158–60.

Brokaw, N. V. L., and J. S. Grear. 1991. Forest structure before and after Hurricane Hugo at three elevations in the Luquillo Mountains, Puerto Rico. *Biotropica* 23: 386–92.

Brown, S., T. Pearson, D. Slaymaker, S. Ambagis, N. Moore, D. Novelo, and W. Sabido. 2005. Creating a virtual tropical forest from three-dimensional aerial imagery to estimate carbon stocks. *Ecological Applications* 15: 1083–95.

Burgess, E., J. Burgess, M. D. Lowman, and thousands of Jason X school students. 2003. Observations of a Coleopteran herbivore on a bromeliad in the Peruvian Amazon. *Journal of the Bromeliad Society* 53: 221–24.

Callaway, R. M., K. O. Reinhart, S. C. Tucker, and S. C. Pennings. 2001. Effects of epiphytic lichens on host preference of the vascular epiphyte *Tillandsia usneoides*. *Oikos* 94: 433–41.

Campbell, R. W., and T. R. Torgersen. 1983. Effect of branch height on predation of western spruce budworm (Lepidoptera: Tortricidae) pupae by birds and ants. *Environmental Entomology* 12: 697–99.

Campbell, R. W., T. R. Torgersen, S. C. Forrest, and L. C. Youngs. 1981. Bird exclosures for branches and whole trees. *USDA Forest Service General Technical Report. PNW-125.* Portland, OR: USDA Forest Service Pacific Northwest Forest and Range Experiment Station.

Campbell, R. W., T. R. Torgersen, and N. Srivastava. 1983. A suggested role for predaceous birds and ants in the population dynamics of the western spruce budworm. *Forest Science* 29: 779–90.

Cardé, R. T., and T. C. Baker. 1984. Sexual communication with pheromones. In *Chemical ecology of insects*, ed. W. J. Bell and R. T. Cardé, 355–83. London: Chapman and Hall.

Cardelús C. L. 2007. Vascular epiphyte communities in the inner-crown of *Hyeronema alchorneoides* and *Lecythis ampla* at La Selva Biological Station, Costa Rica. *Biotropica* 39: 171–76.

Cardelús, C. L., and R. L. Chazdon. 2005. Inner-crown microenvironments of two emergent tree species in a lowland wet forest. *Biotropica* 37: 238–44.

Cardelús, C., R. Colwell, and E. Watkins. 2006. Vascular epiphyte distribution patterns: Explaining the midelevation richness peak. *Journal of Ecology* 94: 144–56.

Cardelús C., M. Lowman, and A. Wassie. 2012. Uniting church and science for conservation. *Science Magazine* 335: 932.

Cardelús, C., and M. Mack. 2010. The nutrient status of epiphytes and their host trees along an elevational gradient in Costa Rica. *Plant Ecology* 207: 25–37.

Cardelús, C. L. 2010. Litter decomposition within the canopy and forest floor of three tree species in a tropical lowland rain forest, Costa Rica. *Biotropica* 42: 300–308.

Carroll, G. C. 1979. Needle microepiphytes in a Douglas fir canopy: Biomass and distribution patterns. *Canadian Journal of Botany* 57: 1000–1007.

———. 1988. Fungal endophytes in stems and leaves: From latent pathogen to mutualistic symbiont. *Ecology* 69: 2–9.

Carter, G. A., and A. K. Knapp. 2001. Leaf optical properties in higher plants: Linking spectral characteristics to stress and chlorophyll concentration. *American Journal of Botany* 88: 677–84.

Cassiani, M., G. G. Katul, and J. D. Albertson. 2008. The effects of canopy leaf area index on airflow across forest edges: Large-eddy simulation and analytical results. *Boundary-Layer Meteorology* 126: 433–60.

Chambers, J. Q., G. P. Asner, D. C. Morton, L. O. Anderson, S. S. Saatchi, F. D. B. Espríto-Santo, M. Palace, and C. Souza Jr. 2007. Regional ecosystem structure and function: Ecological insights from remote sensing of tropical forests. *Trends in Ecology and Evolution* 22: 414–23.

Chapman, S. K., S. C. Hart, N. S. Cobb, T. G. Whitham, and G. W. Koch. 2003. Insect herbivory increases litter quality and decomposition: An extension of the acceleration hypothesis. *Ecology* 84: 2867–76.

Chapman, S. K., J. A. Schweitzer, and T. G. Whitham. 2006. Herbivory differentially alters plant litter dynamics of evergreen and deciduous trees. *Oikos* 114: 566–74.

Chen, J., M. Falk, E. Euskirchen, K. T. Paw U, T. H. Suchanek, S. L. Ustin, B. J. Bond, K. D. Brosofsky, N. Phillips, and R. Bi. 2002. Biophysical controls of carbon flows in three successional Douglas-fir stands based on eddy-covariance measurements. *Tree Physiology* 22: 169–77.

Chen, J., J. F. Franklin, and T. A. Spies. 1995. Growing-season microclimatic gradients from clearcut edges into old-growth Douglas-fir forests. *Ecological Applications* 5: 74–86.

Clark, M. L., D. A. Roberts, and D. B. Clark. 2005. Hyperspectral discrimination of tropical rain forest tree species at leaf to crown scales. *Remote Sensing of Environment* 96: 375–98.

Classen, A. T., S. K. Chapman, T. G. Whitham, S. C. Hart, and G. W. Koch. 2007. Genetic-based plant resistance and susceptibility traits to herbivory influence needle and root litter nutrient dynamics. *Journal of Ecology* 95: 1181–94.

Classen, A. T., S. C. Hart, T. G. Whitham, N. S. Cobb, and G. W. Koch. 2005. Insect infestations linked to changes in microclimate: Important climate change implications. *Soil Science Society of America Journal* 69: 2049–57.

Cobb, N. S., and T. G. Whitham. 1993. Herbivore deme formation on individual trees—a test-case. *Oecologia* 94: 496–502.

Coleman, D. C., D. A. Crossley Jr., and P. F. Hendrix. 2004. *Fundamentals of soil ecology*, 2nd edition. San Diego, CA: Elsevier/Academic Press.

Coley, P. D. 1983. Herbivory and defensive characteristics of tree species in a lowland tropical forest. *Ecological Monographs* 53: 209–33.

Colwell, R. K. 2005. *Estimates: Statistical estimation of species richness and shared species from samples. Version 7.* User's guide and application published at http://viceroy.eeb.uconn.edu/estimates.

Colwell, R., G. Brehm, C. Cardelús, A. Gilman, and J. Longino. 2008. Global warming, elevational range shifts, and lowland biotic attrition in the wet tropics. *Science* 322: 258–61.

Connell, J. H., M. D. Lowman, and I. R. Noble. 1997. Subcanopy gaps in temperate and tropical forests. *Australian Journal of Ecology* 22: 163–68.

Connelly, A. E., and T. D. Schowalter. 1991. Seed losses to feeding by *Leptoglossus occidentalis* (Heteroptera: Coreidae) during two periods of second year cone development in western white pine. *Journal of Economic Entomology* 84: 215–17.

Corbara, B., ed. 2009a. SANTO 2006 Global Biodiversity Survey from sea bottom to ridge crests. *Zoosystema* 31(3): 744.

———. 2009b. Diversité des arthropodes dans une forêt tempérée. *Insectes* 153(2): 3–9.

Cornelissen, J. H. C., S. Lavorel, E. Garnier, S. Diaz, N. Buchmann, D. E. Gurvich, P. B. Reich, et al. 2003. A handbook of protocols for standardised and easy measurement of plant functional traits worldwide. *Australian Journal of Botany* 51: 335–80.

Corner, E. J. H. 1964. *The life of plants.* Chicago: University of Chicago Press.

———. 1992. *Botanical monkeys.* Edinburgh: Pentland Press.

Coxson, D. S., and N. M. Nadkarni. 1995. Ecological roles of epiphytes in nutrient cycles of forest canopies. In *Forest canopies*, ed. M. D. Lowman and N. M. Nadkarni, 395–453. San Diego, CA: Academic Press.

Crossley, D. A., Jr., J. T. Callahan, C. S. Gist, J. R. Maudsley, and J. B. Waide. 1976. Compartmentalization of arthropod communities in forest canopies at Coweeta. *Journal of the Georgia Entomological Society* 11: 44–49.

Crous, K. Y., and D. S. Ellsworth. 2004. Canopy position affects photosynthetic adjustments to long-term elevated CO_2 concentration (FACE) in aging needles in a mature *Pinus taeda* forest. *Tree Physiology* 24: 961–70.

Curran, L. M., S. N. Trigg, A. K. McDonald, D. Astiani, Y. M. Hardiano, P. Siregar, I. Caniago, and E. Kasischke. 2004. Lowland forest loss in protected areas of Indonesian Borneo. *Science* 303: 1000–1003.

Darnaedi, D., and W. A. Noerdjito. 2008. Understanding Indonesian natural diversity: Insect-collecting methods taught to parataxonomists during DIWPA-IBOY training courses. In *Proceedings of international symposium "The Origin and Evolution of Natural Diversity," 1–5 October 2007, Sapporo*, ed. H. Okada, S. F. Mawatari, N. Suzuki, and P. Gautam, 245–50.

Darwin, C. 1859. *The origin of species by means of natural selection.* London: John Murray.

———. 1883. *Journal of researches into the natural history and geology of the countries visited during the voyage of the H.M.S. Beagle round the world.* New York: D. Appleton and Company.

Davidson, E. A., D. C. Nepstad, F. Y. Ishida, and P. M. Brando. 2008. Effects of an experimental drought and recovery on soil emissions of carbon dioxide, methane, nitrous oxide and nitric oxide in a moist tropical forest. *Global Change Biology* 14: 2582–90.

Denison, W. C., D. M. Tracy, F. M. Rhoades, and M. Sherwood. 1972. Direct, nondestructive measurement of biomass and structure in living old-growth Douglas-fir. In *Research on coniferous forest ecosystems*, ed. J. F. Franklin, L. J. Dempster, and R. H. Waring, 147–58. Portland, OR: USDA Forest Service Pacific Northwest Experiment Station.

De Souza, A. L. T., and R. P. Martins. 2005. Foliage density of branches and distribution of plant-dwelling spiders. *Biotropica* 37: 416–20.

Dial, R., B. Bloodworth., A. Lee, P. Boyne, and J. Heys. 2004. The distribution of free space and its relation to canopy composition at six forest sites. *Forest Science* 50: 312–25.

Dial, R., M. D. F. Ellwood, E. C. Turner, and W. A. Foster. 2006. Arthropod abundance, canopy structure, and microclimate in a Bornean lowland tropical rain forest. *Biotropica* 38: 643–52.

Dial, R., and J. Roughgarden. 1995. Experimental removal of insectivores from rain forest canopy: Direct and indirect effects. *Ecology* 76: 1821–34.

———. 2004. Physical transport, hetrogeneity, and interactions involving canopy anoles. In

Forest canopies, 2nd edition, ed. M. D. Lowman and H. B. Rinker, 270–96. San Diego, CA: Elsevier/Academic Press.

Díaz, I. A., K. E. Sieving, M. E. Peña-Foxon, J. Larraín, and J. Armesto. 2010. Epiphyte diversity and biomass loads of canopy emergent trees in Chilean temperate rain forests: A neglected functional component. *Forest Ecology and Management* 259: 1490–501.

Dolch, R., and T. Tscharntke. 2000. Defoliation of alders (*Alnus glutinosa*) affects herbivory by leaf beetles on undamaged neighbors. *Oecologia* 125: 504–11.

Dominy, N. J., P. W. Lucas, and S. J. Wright. 2003. Mechanics and chemistry of rain forest leaves: canopy and understory compared. *Journal of Experimental Botany* 54: 2007–14.

Donati, G., A. Bollen, S. M. Borgognini-Tarli, and J. U. Ganzhorn. 2007. Feeding over the 24-h cycle: Dietary flexibility of cathemeral collared lemurs (*Eulemur collaris*). *Behavioral Ecology and Sociobiology* 61: 1237–51.

Dyer, L. A., and D. K. Letourneau. 1999a. Relative strengths of top-down and bottom-up forces in a tropical forest community. *Oecologia* 119: 265–74.

———. 1999b. Trophic cascades in a complex terrestrial community. *Proceedings of the National Academy of Sciences, USA.* 96: 5072–76.

Edwards, N. T. 1982. The use of soda-lime for measuring respiration rates in terrestrial systems. *Pedobiologia* 23: 321–30.

Ellsworth, D. S., and P. B. Reich. 1993. Canopy structure and vertical patterns of photosynthesis and related leaf traits in a deciduous forest. *Oecologia* 96: 169–78.

Ellwood, M. D. F., D. T. Jones, and W. A. Foster. 2002. Canopy ferns in lowland dipterocarp forest support a prolific abundance of ants, termites, and other invertebrates. *Biotropica* 34: 575–83.

Emmons, L. 1995. Mammals of rain forest canopies. In *Forest canopies*, ed. M. D. Lowman and N. M. Nadkarni, 199–224. San Diego, CA: Academic Press.

Engelmark, O. 1999. Boreal forest disturbances. In *Ecosystems of the world 16: Ecosystems of disturbed ground*, ed. L. R. Walker, 161–86. Amsterdam: Elsevier.

Ernest, K. A., M. D. Lowman, H. B. Rinker, and D. C. Shaw. 2006. Stand-level herbivory in an old-growth conifer forest canopy. *Western North American Naturalist* 66(4): 473–81.

Erwin, T. L. 1982. Tropical forests: Their richness in Coleoptera and other arthropod species. *Coleopterists Bulletin* 36: 74–75.

———. 1983. Tropical forest canopies, the last biotic frontier. *Bulletin Entomological Society America* 29: 14–19.

———. 1990. Canopy arthropod biodiversity: A chronology of sampling techniques and results. *Revista Peruana de Entomologia* 32: 71–77.

———. 1991. How many species are there? Revisited. *Conservation. Biology* 5: 330–33.

Facchini, M. C., M. Mircea, S. Fuzzi, and R. J. Charlson. 1999. Cloud albedo enhancement by surface-active organic solutes in growing droplets. *Nature* 401: 257–59.

Fagan, L. L., R. K. Didham, N. N. Winchester, V. Behan-Pelletier, M. Clayton, E. Lindquist, and R. A. Ring. 2006. An experimental assessment of biodiversity and species turnover in terrestrial vs. canopy leaf litter. *Oecologia* 147: 335–47.

Falge, E., D. Baldocchi, R. Olson, P. Anthoni, M. Aubinet, C. Bernhofer, G. Burba, et al. 2001. Gap filling strategies for long term energy flux data sets. *Agricultural and Forest Meteorology* 107: 71–77.

Fine, P. V. A., I. Mesones, and P. D. Coley. 2004. Herbivores promote habitat specialization by trees in Amazonian forests. *Science* 305: 663–65.

Finnigan, J. 2000. Turbulence in plant canopies. *Annual Review of Fluid Mechanics* 32: 519–71.

Finnigan, J. J., R. H. Shaw and E. G. Patton. 2009. Turbulence structure above a vegetation canopy. *Journal of Fluid Mechanics* 637: 387–424.

Fish, H., V. J. Lieffers, U. Silins, and R. J. Hall. 2006. Crown shyness in lodgepole pine stands of varying stand height, density, and site index in the upper foothills of Alberta. *Canadian Journal of Forest Research* 36: 2104–111.

Floren, A. 2010. Sampling arthropods from the canopy by insecticidal knockdown. In *Manual on field recording techniques and protocols for all taxa biodiversity inventories*, ed. J. Eymann., J. Degreef, Ch. Häuser, J. C. Monje, Y. Samyn, D. Van den Spiegel, ABC Taxa 8 (part 1): 158–72.

Floren, A., and J. Schmidl, ed. 2008. *Canopy arthropod research in Europe: Basic and applied studies from the high frontier.* Nuremberg, Germany: Bioform Entomology and Equipment.

Foley, J. A., M. H. Costa, C. Delire, N. Ramankutty, and P. Snyder. 2003. Green surprise? How terrestrial ecosystems could affect Earth's climate. *Frontiers in Ecology and the Environment* 1: 38–44.

Fonte, S. J., and T. D. Schowalter. 2004a. Decomposition of greenfall vs. senescent foliage in a tropical forest ecosystem in Puerto Rico. *Biotropica* 36: 474–82.

———. 2004b. Decomposition in forest canopies. In *Forest canopies*, 2nd edition, ed. M. D. Lowman and H. B. Rinker, 413–22. San Diego, CA: Elsevier/Academic Press.

———. 2005. The influence of a neotropical herbivore (*Lamponius portoricensis*) on nutrient cycling and soil processes. *Oecologia* 146: 423–31.

Food and Agricultural Organization of the United Nations (FAO). 2001. *State of the world's forests.* Rome: FAO.

Fortin, M., and Y. Mauffette. 2002. The suitability of leaves from different canopy layers for a generalist herbivore (Lepidoptera: Lasiocampidae) foraging on sugar maple. *Canadian Journal of Forest Research* 32: 379–89.

Fox-Dobbs, K., D. F. Doak, A. K. Brody. and T. M. Palmer. 2010. Termites create spatial structure and govern ecosystem function by affecting N_2 fixation in an east African savanna. *Ecology* 91: 1296–307.

Frazer, G. W., R. A. Fournier, J. A. Trofymow, and R. J. Hall. 2001. A comparison of film and digital fisheye photography for analysis of forest canopy structure and gap light transmission. *Agricultural and Forest Meteorology* 109: 249–63.

Frost, C. J., and M. D. Hunter. 2004. Insect canopy herbivory and frass deposition affect soil nutrient dynamics and export in oak mesocosms. *Ecology* 85: 3335–47.

———. 2007. Recycling of nitrogen in herbivore feces: Plant recovery, herbivore assimilation, soil retention, and leaching losses. *Oecologia* 151: 42–53.

———. 2008. Insect herbivores and their frass affect *Quercus rubra* leaf quality and initial stages of subsequent decomposition. *Oikos* 117: 13–22.

Funk, J. L., and M. T. Lerdau. 2004. Photosynthesis in forest canopies. In *Forest canopies*, 2nd edition, ed. M. D. Lowman and H. B. Rinker, 335–58. San Diego, CA: Elsevier/Academic Press.

Furon, A. C., J. S. Warland, and C. Wagner-Riddle. 2007. Analysis of scaling-up resistances from leaf to canopy using numerical simulations. *Journal of Agronomy* 99: 1483–91.

Fyllas, N., S. Patiño, T. Baker, G. B. Nardoto, L. Martinelli, C. Quesada, R. Paiva, et al. 2009. Basin-wide variations in foliar properties of Amazonian forest: phylogeny, soils, and climate. *Biogeosciences* 6: 2677–2708.

Ganesh, T., and S. Devy. 2006. Interactions between nonflying mammals and Cullenia exarillata, a canopy tree from the wet forest of Western Ghats. *Current Science* 90: 1674–79.

Garrity, J. 2010. Evergreen agriculture: A robust approach to sustainable food security in Africa. *Food Security*. doi:10.1007/s12571-010-0700-7.

Gash, J. H. C., and W. J. Shuttleworth. 1991. Tropical deforestation: Albedo and the surface-energy balance. *Climate Change* 19: 123–33.

Gaston, K. 1991. The magnitude of global insect species richness. *Conservation. Biology* 5: 283–96.

Gear, A. J., and B. Huntley. 1991. Rapid changes in the range limits of Scots pine 4000 year ago. *Science* 251: 544–47.

Gehring, C. A., and T. G. Whitham. 1995. Duration of herbivore removal and environmental stress affect the ectomycorrhizae of pinyon pines. *Ecology* 76: 2118–23.

Gering, J. C., K. A. DeRennaux, and T. O. Crist. 2007. Scale dependence of effective specialization: Its analysis and implications for estimates of global insect species richness. *Diversity and Distributions* 13: 115–25.

Gentry, A. H. 1991. The distribution and evolution of climbing plants. In *The biology of vines*, ed. F. E. Putz and H. A. Mooney, 3–42. Cambridge: Cambridge University Press.

Giambelluca, T. W., F. G. Scholz, S. J. Bucci, F. C. Meinzer, G. Goldstein, W. A. Hoffmann, A. C. Franco, and M. P. Buchert. 2009. Evapotranspiration and energy balance of Brazilian savannas with contrasting tree density. *Agricultural and Forest Meteorology* 149: 1365–76.

Gnezdilov, V. M., J. Bonfils, H.-P. Aberlenc, and Y. Basset. 2010. Review of the neotropical genus *Oronoqua* Fennah, 1947 (Insecta, Hemiptera, Issidae, Issinae). *Zoosystema* 32: 247–57.

Göckede, M., T. Foken, and M. Aubinet. 2008. Quality control of CarboEurope flux data. Part I: Coupling footprint analyses with flux data quality assessment to evaluate sites in forest ecosystems. *Biogeosciences* 5: 433–50.

Grace, J. R. 1986. The influence of gypsy moth on the composition and nutrient content of litter fall in a Pennsylvania oak forest. *Forest Science* 32: 855–70.

Granier, A. 1987. Evaluation of transpiration in a Douglas-fir stand by means of sap flow measurements. *Tree Physiology* 3: 309–20.

Greenberg, A. E., L. S. Clesceri, and A. D. Eaton. 1992. *Standard Methods for the Examination of Water and Wastewater*. 18th edition. Washington, DC: American Public Health Association.

Greenstone, M. H. 2006. Molecular methods for assessing insect parasitism. *Bulletin of Entomological Research* 96: 1–13.

Grimbacher, P. S., and N. E. Stork. 2007. Vertical stratification of feeding guilds and body size in beetle assemblages from an Australian tropical rainforest. *Austral Ecology* 32: 77–85.

Grove, S. J. 2002. Saproxylic insect ecology and the sustainable management of forests. *Annual Review of Ecology and Systematics* 33: 1–23.

Guenther, A., J. Greenberg, P. Harley, D. Helmig, L. Klinger, L. Vierling, P. Zimmerman, and

C. Geron. 1996. Leaf, branch, stand and landscape scale measurements of volatile organic compound fluxes from U.S. woodlands. *Tree Physiology* 16: 17–24.

Gunatilleke, N., and S. Gunatilleke. 1996 *Sinharaja: World Heritage Site, Sri Lanka.* Colombo, Sri Lanka: Natural Resources, Energy, and Science Authority (NARESA).

Gutschick, V. P. 1984. Statistical penetration of diffuse light into vegetative canopies: Effect on photosynthetic rate and utility for canopy measurement. *Agricultural Meteorology* 30: 327–41.

———. 1996. Physiological control of evapotranspiration by shrubs: Scaling measurements from leaf to stand with the aid of comprehensive models. In *Proceedings: Shrubland ecosystem dynamics in a changing environment. General. Technical. Report. INT-GTR-338,* ed. J. R. Barrow, E. D. McArthur, R. E. Sosebee, and R. J. Tausch, 214–19. Ogden, UT: USDA Forest Service Intermountain Research Station.

———. 1999. Biotic and abiotic consequences of differences in leaf structure. *New Phytologist* 143: 4–18.

Gutschick, V. P., and F. W. Wiegel. 1988. Optimizing the canopy photosynthetic rate by patterns of investment in specific leaf mass. *American Naturalist* 132: 67–86.

Haddow, A. J., P. S. Corbet, and J. D. Gilett. 1961. Studies from a high tower in Mpanga Forest, Uganda. *Transactions of the Royal Entomological Society of London* 113: 249–368.

Hallé, F. 1990. A raft atop the rain forest. *National Geographic* 178: 129–38.

Hallé, F., and P. Blanc, ed. 1990. *Biologie d'une canopée de forêt equatorial.* Montpelier, France: Institut Botanique.

Hallé, F., R. A. A. Oldeman, and P. B. Tomlinson. 1978. *Tropical trees and forests—An architectural analysis.* New York: Springer-Verlag.

Hallé, F., and O. Pascal, eds. 1992. *Biologie d'une canopée de forêt equatoiale. II. Rapport de mission: Radeau des cimes Octobre/Novembre 1991.* Paris: Foundation Elf.

Hamilton, A. J., Y. Basset, K. K. Benke, P. S. Grimbacher, S. E. Miller, V. Novotny, G. A. Samuelson, N. Stork, G. D. Weiblen, and J. D. L. Yen. 2010. Quantifying uncertainty in estimation of global arthropod species richness. *American Naturalist* 176: 90–95.

Hargrove, W. W. 1988. A photographic technique for tracking herbivory on individual leaves through time. *Ecological Entomology* 13: 359–63.

Hargrove, W. W., D. A. Crossley Jr., and T. R. Seastedt. 1984. Shifts in insect herbivory in the canopy of black locust, *Robinia pseudoacacia,* after fertilization. *Oikos* 43: 322–28.

Harley, P., A. Guenther, and P. Zimmerman. 1996. Effects of light, temperature, and canopy position on net photosynthesis and isoprene emission from sweetgum (*Liquidambar styraciflua*) leaves. *Tree Physiology* 16: 25–32.

Harley, P., A. Guenther, and P. Zimmerman. 1997. Environmental controls over isoprene emission in deciduous oak canopies. *Tree Physiology* 17: 705–14.

Hättenschwiler, S., and C. Schafellner. 2004. Gypsy moth feeding in the canopy of a CO_2-enriched mature forest. *Global Change Biology* 10: 1899–908.

Hawken, P., A. Lovins, and L. H. Lovins. 1999. *Natural capitalism.* New York: Little, Brown and Co.

Head, S., and R. Heinzman. 1990. *Lessons of the rainforest.* Washington, DC: Island Press.

Heald, C. L., M. J. Wilkinson, R. K. Monson, C. A. Alo, G. Wang, and A. Guenther. 2009. Response of isoprene emission to ambient CO_2 changes and implications for global budgets. *Global Change Biology* 15: 1127–40.

Heartsill-Scalley, T., F. N. Scatena, C. Estrada, W. H. McDowell, and A. E. Lugo. 2007. Disturbance and long-term patterns of rainfall and throughfall nutrient fluxes in a subtropical wet forest in Puerto Rico. *Journal of Hydrology* 333: 472–85.

Heatwole, H. 1989a. Changes in ant assemblages across an arctic treeline. *Revue d'Entomologie dur Quebec* 34: 10–22.

———. 1989b. The concept of the econe, a fundamental ecological unit. *Tropical Ecology* 30(1): 13–19.

Hebert, P. D. N., E. H. Penton, J. M. Burns, D. H. Janzen, and W. Hallwachs. 2004. Ten species in one: DNA barcoding reveals cryptic species in the neotropical skipper butterfly Astraptes fulgerator. *Proceedings of the National Academy of Sciences, USA* 101: 14812–17.

Heinsch, F. A., M. Zhao, S. W. Running, J. S. Kimball, R. R. Nemani, K. J. Davis, et al. 2006. Evaluation of remote sensing based terrestrial productivity from MODIS using regional tower eddy flux network observations. *IEEE Transactions on Geoscience and Remote Sensing* 44: 1908–25.

Helmer, E. H., T. S. Ruzycki, J. M. Wunderle Jr., S. Vegesser, B. Ruefenacht, C. Kwit, T. J. Brandeis, and D. N. Ewert. 2010. Mapping tropical dry forest height, foliage height profiles and disturbance type and age with a time series of cloud-cleared Landsat and ALI image mosaics to characterize avian habitat. *Remote Sensing of Environment* 114: 2457–73.

Hendry, G. R., and B. A. Kimball. 1994. The FACE program. *Agricultural and Forest Meteorology* 70: 3–14.

Hendry, G. R., D. S. Ellsworth, K. F. Lewin, and J. Nagy. 1999. A free-air enrichment system for exposing tall forest vegetation to elevated atmospheric CO_2. *Global Change Biology* 5: 293–309.

Herrick, J. D., and R. B. Thomas. 2003. Leaf senescence and late-season net photosynthesis of sun and shade leaves of overstory sweetgum (*Liquidambar styraciflua*) grown in elevated and ambient carbon dioxide concentrations. *Tree Physiology* 23: 109–18.

Hietz, P. and J. Ausserer. 1998. Population dynamics and growth of epiphytes in a humid montane forest in Mexico. *Selbyana* 19: 281.

Hingston, R. W. G. 1932. *A naturalist in the Guiana forest*. New York: Longmans, Green.

Hirao, T., M. Murakami, A. Kashizaki, and S.-I. Tanabe. 2007. Additive apportioning of lepidopteran and coleopteran species diversity across spatial and temporal scales in a co-ol-temperate deciduous forest in Japan. *Ecological Entomology* 32: 627–36.

Holdridge, L. R. 1970. A system for representing structure in tropical forest associations. In *A tropical rain forest*, ed. H. T. Odum and R. F. Pigeon, B-147–50. Oak Ridge, TN: US Atomic Energy Commission.

Holland, J. N., W. Cheng, and D. A. Crossley Jr. 1996. Herbivore-induced changes in plant carbon allocation: Assessment of below-ground C fluxes using carbon-14. *Oecologia* 107: 87–94.

Hollinger, D. Y. 1986. Herbivory and the cycling of nitrogen and phosphorus in isolated California oak trees. *Oecologia* 70: 291–97.

Horn, H. S. 1971. *The adaptive geometry of trees*. Princeton, NJ: Princeton University Press.

Houlahan, J. E., C. S. Findlay, B. R. Schmidt, A. H. Meyer, and S. L. Kuzmin. 2000. Quantitative evidence for global amphibian population declines. *Nature* 404: 752–55.

Hubbell, S. P., and R. B. Foster. 1983. Diversity of canopy trees in a neotropical forest and implications for conservation. *Tropical rain forest: Ecology and management,* ed. T. C. Whitmore, A. C. Chadwick, and A. C. Sutton, 25–41. Oxford: British Ecological Society.

Hunter, M. D. 2001. Insect population dynamics meets ecosystem ecology: Effects of herbivory on soil nutrient dynamics. *Agricultural and Forest Entomology* 3: 77–84.

Hurlbert, S. H. 1984. Pseudoreplication and the design of ecological field experiments. *Ecological Monographs* 54: 187–211.

Ingham, E. R. 1985. Review of the effects of 12 selected biocides on target and non-target soil organisms. *Crop Protection* 4: 3–32.

Irvine, J., B. E. Law, P. M. Antoni, and F. C. Meinzer. 2002. Water limitations to carbon exchange in old-growth and young ponderosa pine stands. *Tree Physiology* 22: 189–96.

Irvine, J., B. E. Law, and M. R. Kurpius. 2005. Coupling of canopy gas exchange with root and rhizosphere respiration in a semiarid forest. *Biogeochemistry* 73: 271–82.

Irvine, J., B. E. Law, J. Martin, D. Vickers. 2008. Interannual variation in soil CO_2 efflux and the response of root respiration to climate and canopy gas exchange in mature ponderosa pine. *Global Change Biology* 14: 2848–59.

Ishii, H., and M. E. Wilson. 2001. Crown structure of old-growth Douglas-fir in the western Cascade Range, Washington. *Canadian Journal of Forest Research* 31: 1250–61.

Jactel, H., and E. G. Brockerhoff. 2007. Tree diversity reduces herbivory by forest insects. *Ecology Letters* 10: 835–58.

Jansen, P. A., S. Bohlman, C. X. Garzon-Lopez, H. Olff, H. C. Muller-Landau, and S. J. Wright. 2008. Large-scale spatial variation in palm fruit abundance across a tropical moist forest estimated from high-resolution aerial photographs. *Ecography* 31: 33–42.

Janssen, R. H. H., M. B. J. Meinders, E. H. van Nes, and M. Scheffer. 2008. Microscale vegetation-soil feedback boosts hysteresis in a regional vegetation-climate system. *Global Change Biology* 14: 1104–12.

Janzen, D. H. 1992. A south–north perspective on science in the management, use, and economic development of Parataxonomists in Costa Rica biodiversity. In *Conservation of Biodiversity for Sustainable Development,* ed. O. T. Sandlund, K. Hindar, and A. H. D. Brown, 27–52. Oslo, Norway: Scandinavian University Press.

———. 2004. Setting up tropical biodiversity for conservation through nondamaging use: Participation by parataxonomists. *Journal of Applied Ecology* 41: 181–87.

———. 2010. Hope for tropical biodiversity through true bioliteracy. *Biotropica* 42(5): 540–42.

Jarzen, D., S. A. Jarzen, and M. D. Lowman 2010. In and out of Africa. *The Palynological Society Newsletter* 43(4): 11–15.

Jenkins, M. J., E. Herbertson, W. Page, and C. A. Jorgensen. 2008. Bark beetles, fuels, fires and implications for forest management in the intermountain west. *Forest Ecology and Management* 254: 16–34.

Juang, J.-Y., G. G. Katul, A. Porporato, P. C. Stoy, M. S. Sequeira, M. Detto, H.-S. Kim, and R. Oren. 2007. Eco-hydrological controls on summertime convective rainfall triggers. *Global Change Biology* 13: 887–96.

Julliot, C. 1997. Impact of seed dispersal by red howler monkeys *Allouatta seniculus* on the seedling population in the understory of tropical rain forest. *Journal of Ecology* 85: 431–40.

Kays, R., and A. Allison. 2001. Arboreal tropical forest vertebrates: Current knowledge and research trends. *Plant Ecology* 153: 109–20.

Kimmins, J. P. 1972. Relative contributions of leaching, litterfall, and defoliation by *Neodiprion sertifer* (Hymenoptera) to the removal of cesium–134 from red pine. *Oikos* 23: 226–34.

Kimura, S. D., M. Saito, H. Hara, Y. H. Xu, and M. Okazaki. 2009. Comparison of nitrogen dry deposition on cedar and oak leaves in the Tama Hills using foliar rinsing method. *Water, Air, and Soil Pollution* 202: 369–77.

Kinchin, I. 1994. *The biology of tardigrades*. London: Portland Press.

King, R. A., D. S. Read, M. Traugott, and W. O. C. Symondson. 2008. Molecular analysis of predation: A review of best practice for DNA-based approaches. *Molecular Ecology* 17: 947–63.

Knepp, R. G., J. G. Hamilton, J. E. Mohan, A. R. Zangerl, M. R. Berenbaum, and E. H. DeLucia. 2005. Elevated CO_2 reduces leaf damage by insect herbivores in a forest community. *New Phytologist* 167: 207–18.

Koch, G. W., S. C. Sillett, G. M. Jennings, and S. D. Davis. 2004. The limits to tree height. *Nature* 28: 851–54.

Kolb, T. E., K. A. Dodds, and K. M. Clancy. 1999. Effect of western spruce budworm defoliation on the physiology and growth of potted Douglas-fir seedlings. *Forest Science* 45: 280–91.

Körner, C., R. Asshoff, O. Bignucolo, S. Hättenschwiler, S. G. Keel, S. Peláez-Riedl, S. Pepin, R. T. W. Siegwolf, and G. Zotz. 2005. Carbon flux and growth in mature deciduous forest trees exposed to elevated CO_2. *Science* 309: 1360–62.

Krömer, T., M. Kessler, and S. R. Gradstein. 2007. Vertical stratification of vascular epiphytes in submontane and montane forest of the Bolivian Andes: The importance of the understory. *Plant Ecology* 189: 261–78.

Kuuluvainen, T. 1992. Tree architectures adapted to efficient light utilization: Is there a basis for latitudinal gradients? *Oikos* 65: 275–84.

Landsberg, J., and C. P. Ohmart. 1989. Levels of defoliation in forests: Patterns and concepts. *Trends in Ecology and Evolution* 4: 96–100.

Laurance, W. F., and C. A. Perez, eds. 2006. *Emerging threats to tropical forests*. Chicago: University of Chicago Press.

Laurence, W. F., K. R. McDonald, and R. Speare. 1996. Epidemic disease and the catastrophic decline of Australian rain forest frogs. *Conservation Biology* 10(2): 406–13.

Law, B. E. 2005. Carbon dynamics in response to climate and disturbance: Recent progress from multiscale measurements and modeling in AmeriFlux. In *Plant responses to air pollution and global change*, ed. K. Omasa, I. Nouchi, and L. J. de Kok, 1–9. Tokyo: Springer-Verlag.

Law, B. E., E. Falge, D. D. Baldocchi, P. Bakwin, P. Berbigier, K. Davis, A. J. Dolman, M., et al. 2002. Environmental controls over carbon dioxide and water vapor exchange of terrestrial vegetation. *Agricultural and Forest Meteorology* 113: 97–120.

Law, B. E., D. Turner, J. Campbell, M. Lefsky, M. Guzy, O. Sun, S. Van Tuyl, and W. Cohen. 2006. Carbon fluxes across regions: Observational constraints at multiple scales. In *Scaling and uncertainty analysis in ecology: Methods and applications*, ed. J. Wu, K. B. Jones, H. Li, and O. L. Loucks, 167–90. Amsterdam: Springer.

Law, B. E., D. Turner, J. Campbell, O. J. Sun, S. Van Tuyl, W. D. Ritts, and W. B. Cohen. 2004. Disturbance and climate effects on carbon stocks and fluxes across western Oregon USA. *Global Change Biology* 10: 1429–44.

Lawton, J. H., D. E. Bignell, B. Bolton, G. F. Bloemers, P. Eggleton, P. M. Hammond, M. Hodda, et al. 1998. Biodiversity inventories, indicator taxa and effects of habitat modification in tropical forest. *Nature* 391: 72–76.

Leather, S. R., ed. 2005. *Insect sampling in forest ecosystems*. Malden, MA: Blackwell Science.

Lefsky, M. A. 2010. A global forest canopy height map from the moderate resolution imaging spectroradiometer and the geoscience laser altimeter system. *Geophysical Research Letters* 37, L15401. doi:10.1029/2010GL043622.

Lefsky, M. A., A. T. Hudak, W. B. Cohen, and S. A. Acker. 2005. Geographic variability in lidar predictions of forest stand structure in the Pacific Northwest. *Remote Sensing of Environment* 95: 532–48.

Lelieveld, J., T. M. Butler, J. N. Crowley, T. J. Dillon, H. Fischer, L. Ganzeveld, H. Harder, M. G. Lawrence, M. Martinez, D. Taraborrelli, and J. Williams. 2008. Atmospheric oxidation capacity sustained by a tropical forest. *Nature* 452: 737–40.

Leopold, A. 1949. *Sand county almanac*. Oxford: Oxford University Press.

Leponce, M., C. Meyer, C. Häuser, P. Bouchet, J. H. C. Delabie, L. Weigt, and Y. Basset. 2010. Challenges and solutions for planning and implementing large-scale biotic inventories. In *Manual on field recording techniques and protocols for all taxa biodiversity inventories*, ed. J. Eymann., J. Degreef, Ch. Häuser, J. C. Monje, Y. Samyn, D. Van den Spiegel, ABC Taxa 8: 19–49.

Lerdau, M. T., A. Guenther, and R. Monson. 1997. Plant production and emission of volatile organic compounds. *BioScience* 47: 373–83.

Lerdau, M. T., and H. L. Throop. 1999. Isopene emission and photosynthesis in a tropical forest canopy: Implications for model development. *Ecological Applications* 9: 1109–17.

Leuning, R., S. J. Zegelin, K. Jones, H. Keith, and D. Hughes. 2008. Measurement of horizontal and vertical advection of CO_2 within a forest canopy. *Agricultural and Forest Meteorology* 148: 1777–97.

Leuzinger, S., and C. Körner. 2007. Water savings in mature deciduous forest trees under elevated CO_2. *Global Change Biology* 13: 2498–508.

Lewis, T. 1998. The effect of deforestation on ground surface temperatures. *Global and Planetary Change* 18: 1–13.

Lindo, Z., and N. N. Winchester. 2007. Oribatid mite communities and foliar litter decomposition in canopy suspended soils and forest floor habitats of western red cedar forests, Vancouver Island, Canada. *Soil Biology and Biochemistry* 39: 2957–66.

———. 2008. Scale dependent diversity patterns in arboreal and terrestrial oribatid mite (Acari: Oribatida) communities. *Ecography* 31: 53–60.

———. 2009. Spatial and environmental factors contributing to patterns in arboreal and terrestrial oribatid mite diversity across spatial scales. *Oecologia* 160: 817–25.

Lindo, Z., N. N. Winchester, and R. K. Didham. 2008. Nested patterns of community assembly in the colonisation of artificial canopy habitats by oribatid mites. *Oikos* 117: 1856–64.

Lindroth, R. L., B. J. Kopper, W. F. J. Parsons, J. G. Bockheim, D. F. Karnosky, G. R. Hendrey, K. S. Pregitzer, J. G. Isebrands, and J. Sober. 2001. Consequences of elevated carbon dioxide and ozone for foliar chemical composition and dynamics in trembling aspen (*Populus tremuloides*) and paper birch (*Betula papyrifera*). *Environmental Pollution* 115: 395–404.

Linsenmair, K. E., A. J. Davis, B. Fiala, M. R. Speight, eds. 2001. *Tropical forest canopies: Ecology and management*. Dordrecht, Netherlands: Kluwer Academic.

Lips, K. R. 1998. Decline of a tropical montane amphibian fauna. *Conservation Biology* 12(1): 106–17.

Longino, J. T., and R. K. Colwell. 1997. Biodiversity assessment using structured inventory: Capturing the ant fauna of a tropical rain forest. *Ecological Applications* 7: 1263–77.

Louv, R. 2005. *Last child in the woods*. Chapel Hill, NC: Algonquin Press.

Lovejoy, T. E. 1995. Foreword. In *Forest canopies*, ed. M. D. Lowman and N. M. Nadkarni, xv–xvi. San Diego, CA: Academic Press.

Lovejoy, T. E., and R. O. Bierregaard. 1990. Central Amazonian forests and the minimum critical size of ecosystems project. In *Four neotropical rainforests*, ed. A. H. Gentry, 60–71. New Haven CT: Yale University Press.

Lovejoy, T. E., and L. Hannah, eds. 2005. *Climate change and biodiversity*. New Haven, CT: Yale University Press.

Lovett, G. M., and S. E. Lindberg. 1993. Atmospheric deposition and canopy interactions of nitrogen in forests. *Canadian Journal of Forest Research* 23: 1603–16.

Lowman, M. D. 1978. Phenology and productivity of *Betula pendula* and *B. pubescens* in Scotland. Master's thesis, Aberdeen University.

———. 1984. An assessment of techniques for measuring herbivory: Is rainforest defoliation more intense than we thought? *Biotropica* 16(4): 264–68.

———. 1985. Spatial and temporal variability in herbivory of Australian rain forest canopies. *Australian Journal of Ecology* 10: 7–14.

———. 1986. Light interception and its relation to structural differences in three Australian rainforest canopies. *Australian Journal of Ecology* 11: 163–70.

———. 1988. Litterfall and leaf decay in three Australian biotropica rainforest formations. *Journal of Ecology* 76: 51–465.

———. 1992. Leaf growth dynamics and herbivory in five species of Australian rainforest canopy trees. *Journal of Ecology* 80: 433–47.

———. 1995. Herbivory as a canopy process in rain forest trees. In *Forest canopies*, ed. M. D. Lowman and N. M. Nadkarni, 431–561. San Diego, CA: Academic Press.

———. 1997. Herbivory in forests-from centimetres to megametres. In *Forests and insects*, ed. A. D. Watt, N. E. Stork, and M. D. Hunter, 136–49. Boca Raton, FL: Chapman and Hall.

———. 1998. Artificial bromeliad tank experiments. From the Jason X program curriculum: 189–91.

———. 1999. *Life in the treetops*. New Haven, CT: Yale University Press.

———. 2001. Plants in the forest canopy: Some reflections on current research and future direction. *Plant Ecology* 153: 39–50.

———. 2004a. Tarzan or Jane? A short history of canopy biology. In *Forest canopies*, 2nd edition, ed. M. D. Lowman and H. B. Rinker, 453–65. San Diego, CA: Elsevier/Academic Press.

———. 2004b. Ecotourism and the treetops. In *Forest canopies*, 2nd edition, ed. M. D. Lowman and H. B. Rinker, 465–74. San Diego, CA: Elsevier/Academic Press.

———. 2009a. Canopy research in the twenty-first century: A review of arboreal ecology. *Tropical Ecology* 50: 125–36.

———. 2009b. Canopy walkways for conservation: A tropical biologist's panacea or fuzzy metrics to justify ecotourism. *Biotropica* 41(5): 545–48.

———. 2011. Finding sanctuary—saving the biodiversity of Ethiopia, one church forest at a time. *The Explorers Journal* (Winter): 26–31.

Lowman, M. D., and B. Bouricius. 1995. The construction of platforms and bridges for forest canopy access. *Selbyana* 16: 179–84.

Lowman, M. D., E. Burgess, and J. Burgess. 2006. *It's a jungle up there.* New Haven, CT: Yale University Press.

Lowman, M. D., C. D'Avanzo, and C. Brewer. 2009 A national ecological network for research and education. *Science* 323: 1176–77.

Lowman, M. D., and H. H. Heatwole. 1992. Spatial and temporal variability in defoliation of Australian eucalypts. *Ecology* 73: 129–42.

Lowman, M. D., R. L. Kitching, and G. Carruthers. 1996. Arthropod sampling in Australian subtropical rain forest: How accurate are some of the more common techniques? *Selbyana* 17: 36–42.

Lowman, M. D., and M. Moffett. 1993. The ecology of tropical rain forest canopies. *Trends in Ecology and Evolution* xx: 104–7.

Lowman, M. D., M. Moffett, and H. B. Rinker. 1993. A new technique for taxonomic and ecological sampling in rain forest canopies. *Selbyana* 14: 75–79.

Lowman, M. D., and T. Mourad. 2010. Bridging the divide between virtual and real nature. *Frontiers in Ecology and the Environment* 8(7): 339.

Lowman, M. D., and H. B. Rinker, eds. 2004. *Forest canopies.* 2nd edition. San Diego, CA: Elsevier/Academic Press.

Lowman, M. D., and B. J. Selman. 1983. The biology and herbivory rates of *Novacastria nothofagi* Selman (Coleoptera: Chrysomelidae), a new genus and species on *Nothofagus moorei* in Australian temperate rain forests. *Australian Journal of Zoology* 31: 179–91.

Lowman, M. D., and P. K. Wittman. 1996. Forest canopies: Methods, hypotheses, and future directions. *Annual Review of Ecology and Systematics* 27: 55–81.

Luyssaert, E.-D., A. Schulze, A. Börner, A. Knohl, D. Hessenmöller, B. E. Law, P. Ciais, and J. Grace. 2008. Old-growth forests as global carbon sinks. *Nature* 455: 213–15. doi:10.1038.

Lyons, B., N. M. Nadkarni, and M. P. North. 2000. Spatial distribution and succession of epiphytes on *Tsuga heterophylla* (western hemlock) in an old-growth Douglas-fir forest. *Canadian Journal of Botany* 78: 957–68.

Madigosky, S. R. 2004. Tropical microclimatic considerations. In *Forest canopies*, 2nd edition, ed. M. D. Lowman and H. B. Rinker, 24–48. San Diego, CA: Elsevier/Academic Press.

Mafra-Neto, A., and R. T. Cardé. 1995. Influence of plume structure and pheromone concentration on upwind flight by *Cadra cautella* males. *Physiological Entomology* 20: 117–33.

Majer, J. D., and H. F. Recher. 1988. Invertebrate communities on Western Australian eucalypts—A comparison of branch clipping and chemical knockdown procedures. *Australian Journal of Ecology* 13: 269–78.

Malcolm J. R. 1991. Comparative abundances of neotropical small mammals by trap height. *J. Mammal.* 72: 188–202.

Malcolm, J. R. 1995. Forest structure and the abundance and diversity of Neotropical small mammals. In *Forest canopies*, ed. M. D. Lowman and N. M. Nadkarni, 179–223. San Diego, CA: Academic Press.

Malcolm, J. R. 2004. Ecology and conservation of canopy mammals. In *Forest canopies*, 2nd edition, ed. M. D. Lowman and H. B. Rinker, 297–331. San Diego, CA: Elsevier/Academic Press.

Mariotto, I., and V. P. Gutschick. 2010. Non-Lambertian corrected albedo and vegetation index for estimating land evapotranspiration in a heterogeneous semiarid landscape. *Remote Sensing* 2: 926–38.

Marquis, R. J., and C. J. Whelan. 1994. Insectivorous birds increase growth of white oak through consumption of leaf-chewing insects. *Ecology* 75: 2007–14.

Marschner, H. 1995. *The mineral nutrition of higher plants*. 2nd edition. San Diego, CA: Academic Press.

Massman, W. J., 2000. A simple method for estimating frequency response corrections for eddy covariance systems. *Agricultural and Forest Meteorology* 104: 185–98.

Massman, W., X. Lee, and B. E. Law, eds. 2004. *Handbook of micrometeorology. A guide for surface flux measurements and analysis*. Boston: Kluwer Academic.

Matthews, E., and A. Hammond. 1999. *Critical consumption trends and implications: Degrading Earth's resources*. Washington, DC: World Resources Institute.

Mattson, W. J., and R. A. Haack. 1987. The role of drought in outbreaks of plant-eating insects. *BioScience* 37: 110–18.

May, R. 2010. Tropical arthropod species, more or less? *Science* 239: 41–42.

McCarthy, H. R., R. Oren, K. H. Johnsen, A. Gallet-Budynek, S. G. Pritchard, C. W. Cook, S. L. LaDeau, R. B. Jackson, and A. C. Finzi. 2010. Reassessment of plant carbon dynamics at the Duke free-air CO_2 enrichment site: Interactions of atmospheric $[CO_2]$ with nitrogen and water availability over stand development. *New Phytologist* 185: 514–28.

McClure, H. E. 1966. Flowering, fruiting and animals in the canopy of a tropical rain forest. *Malaysian Forester* 29: 182–203.

McCullough, D. G., R. A. Werner, and D. Neumann. 1998. Fire and insects in northern and boreal forest ecosystems of North America. *Annual Review of Entomology* 43: 107–27.

McDowell, W. H. 1998. Internal nutrient fluxes in a tropical rain forest. *Journal of Tropical Ecology* 14: 521–36.

McNaughton, S. J. 1985. Ecology of a grazing system: The Serengeti. *Ecological Monographs* 55: 259–94.

———. 1986. On plants and herbivores. *American Naturalist* 128: 765–70.

———. 1993. Grasses and grazers, science and management. *Ecological Applications* 3: 17–20.

Meadows, D. 1987. *The global citizen*. Washington, DC: Island Press.

Means, J. E., S. A. Acker, D. J. Hardin, J. B. Blair, M. A. Lefsky, W. B. Cohen, M. E. Harmon, and W. A. McKee. 1999. Use of large-footprint scanning airborne LiDAR to estimate forest stand characteristics in the western cascades of Oregon. *Remote Sensing of Environment* 67: 298–308.

Meentemeyer, V. 1978. Macroclimate and lignin control of litter decomposition rates. *Ecology* 59: 465–72.

Meher-Homji, V. M. 1991. Probable impact of deforestation on hydrological processes. *Climate Change* 19: 163–73.

Mehra, P. N., and K. S. Bawa. 1968. B-Chromosomes in some Himalayan hardwoods. *Chromosoma* 25: 90–95.

Mehra, P. N., and K. S. Bawa. 1969. Chromosomal evolution in tropical hardwoods. *Evolution* 23: 466–81.

Melillo, J. M., P. A. Steudler, J. D. Aber, K. Newkirk, H. Lux, F. P. Bowles, C. Catricala, A Magill, T. Ahrens, and S. Morrisseau. 2002. Soil warming and carbon cycle feedbacks to the climate system. *Science* 298: 2173–76.

Mihailovic, D. T., K. Alapati, and Z. Podrascanin. 2009. Chemical transport models: The combined nonlocal diffusion and mixing schemes, and calculation of in-canopy resistance for dry deposition fluxes. *Environmental Science and Pollution Research* 16: 144–51.

Millenium Ecosystem Assessment. 2005. *Ecosystems and human well-being: Biodiversity synthesis.* Washington, DC: World Resources Institute.

Miller, K. K., and M. R. Wagner. 1984. Factors influencing pupal distribution of the Pandora moth (Lepidoptera: Saturniidae) and their relationship to prescribed burning. *Environmental Entomology* 13: 430–31.

Miller, S. E. 2007. DNA barcoding and the renaissance of taxonomy. *Proceedings of the National Academy of Sciences, USA* 104: 4775–76.

Miller, W. 2004. Tardigrades. In *Forest canopies,* 2nd edition, ed. M. D. Lowman and H. B. Rinker, 251–58. San Diego, CA: Elsevier/Academic Press.

Millican, A. 1891. *The travels and adventures of an orchid hunter.* London: Castle and Company.

Misson, L., D. D. Baldocchi, T. A. Black, P. D. Blanken, Y. Brunet, J. Curiel Yuste, J. R. Dorsey, et al. 2007. Partitioning forest carbon fluxes with overstory and understory eddy-covariance measurements: A synthesis based on FLUXNET data. *Agricultural and Forest Meteorology* 144: 14–31.

Mitchell A. 1982. *Reaching the rainforest roof—A handbook on techniques of access and study in the canopy.* Leeds, UK: *Leeds Philosophical and Literary Society/UNEP.*

———. 2001. Canopy science: Time to shape up. In *Tropical forest canopies: Ecology and management,* ed. K. E. Linsenmair, A. J. David, B. Fiala, and M. R. Speight, 5–13. Dordrecht, Netherlands: Kluwer Academic.

Mitchell, A. W. 1986. *The enchanted canopy.* Glasgow, Scotland: Williams Collins and Sons.

Mitchell, A. W., K. Secoy, and T. Jackson, eds. 2002. *The global canopy handbook.* Oxford, UK: Global Canopy Programme.

Mitchell, C., W. R. Miller, and B. Davis. 2009. Tardigrades of North America: Influence of substrate on habitat selection. *Journal of the Pennsylvania Academy of Science* 83(1): 10–16.

Mizutani, M., and N. Hijii. 2001. Mensuration of frass drop for evaluating arthropod biomass in canopies: A comparison among *Cryptomeria japonica, Larix kaempferi,* and deciduous broad-leaved trees. *Forest Ecology and Management* 154: 327–35.

Moffett, M. W. 1993. *The high frontier.* Cambridge, MA: Harvard University Press.

———. 2000. What's "up"? A critical look at the basic terms of canopy biology. *Biotropica* 32: 569–96.

Moffett, M. W., and M. D. Lowman. 1995. Canopy access techniques. In *Forest canopies,* ed. M. D. Lowman and N. M. Nadkarni, 3–26. San Diego, CA: Academic Press.

Moldenke, A. R. 1976. California pollination ecology and vegetation types. *Phytologia* 34: 305–61.

Monteith, J. L. 1973. *Principles of environmental physics.* New York: American Elsevier.

Mooney, K. A. 2007. Tritrophic effects of birds and ants on a canopy food web, tree growth, and phytochemistry. *Ecology* 88: 2005–14.

Moore, J. R., and D. A. Maguire. 2005. Natural sway frequencies and damping ratios of trees: Influence of crown structure. *Trees* 19: 363–73.

Moran, M. D. 2003. Arguments for rejecting the sequential Bonferroni in ecological studies. *Oikos* 100: 403–5.

Mori, S. A. 1984. Use of Swiss tree grippers for making botanical collections of tropical trees. *Biotropica* 16: 79–80.

Mulkey, S. S., R. L. Chazdon, and A. P. Smith, eds. 1996. *Tropical forest plant ecophysiology.* New York: Chapman and Hall.

Muller-Schwarze, D. 2009. *Hands-on chemical ecology.* New York: Springer-Verlag.

Munn, C. A., and B. A. Loiselle. 1995. Canopy access techniques and their importance for the study of tropical forest canopy birds. In *Forest canopies,* ed. M. D. Lowman and N. M. Nadkarni, 165–78. San Diego, CA: Academic Press.

Murawski, D. A. 1995. Reproductive biology and genetics of tropical trees from a canopy perspective. In *Forest canopies,* ed. M. D. Lowman and N. M. Nadkarni, 457–94. San Diego, CA: Academic Press.

Muraoka, H., and H. Koizumi. 2009. Satellite ecology (SATECO)—linking ecology, remote sensing and micrometeorology, from plot to regional scale, for the study of ecosystem structure and function. *Journal of Plant Research* 122: 3–20.

Murlis, J., J. S. Elkinton, and R. T. Cardé. 1992. Odor plumes and how insects use them. *Annual Review of Entomology* 37: 505–32.

Murray, S. N., R. F. Ambrose, and M. N. Dethier. 2006. *Monitoring rocky shores.* Berkeley, CA: University of California Press.

Muul, I., and L. B. Liat. 1970. Vertical zonation in a tropical forest in Malaysia: Methods of study. *Science* 196: 788–89.

Myers, M. 1999. The world's forests and their ecosystem services. In *Nature's services: Societal dependence on natural ecosystems,* ed. G. C. Daily, 215–35. Washington, DC: Island Press.

Nadkarni, N. 1984. Epiphyte biomass and nutrient capital of a neotropical elfin forest. *Biotropica* 16: 249–56.

———. 1995. Good-bye, Tarzan. *The Sciences* 35(1): 28–33.

———. 2008. *Between earth and sky.* San Diego CA: University of California Press.

Nadkarni, N., and T. Matelson. 1991. Fine litter dynamics within the tree canopy of a tropical cloud forest. *Ecology* 72: 2071–82.

Nadkarni, N. M., G. G. Parker, H. B. Rinker, and D. M. Jarzen. 2004. The nature of forest canopies. In *Forest canopies,* 2nd edition, ed. M. D. Lowman and H. B. Rinker, 3–23. San Diego, CA: Elsevier/Academic Press.

Nakadai, T., H. Koizumi, Y. Usami, M. Satoh, and T. Oikawa. 1993. Examination of the method for measuring soil respiration in cultivated land: Effect of carbon dioxide concentration on soil respiration. *Ecological Research* 8: 65–71.

Nansen, C., T. Macedo, R. Swanson, and D. K. Weaver. 2009. Use of spatial structure analysis hyperspectral data cubes for detection of insect-induced stress in wheat plants. *International Journal of Remote Sensing* 30: 2447–64.

Nansen, C., A. J. Sidumo, and S. Capareda. 2010. Variogram analysis of hyperspectral data to characterize the impact of biotic and abiotic stress of maize plants and to estimate biofuel potential. *Applied Spectroscopy* 64: 627–36.

National Research Council (NRC). 2010. *Verifying greenhouse gas emissions: Methods to support international climate agreements.* Washington, DC: National Academies Press.

Nelson, P., J. Nelson, and D. Larkin 2000. *The treehouse book.* New York: Rizzoli International.

Nepstad, D. C., I. M. Tohver, D. Ray, P. Moutinho, and G. Cardinot. 2007. Mortality of large trees and lianas following experimental drought in an Amazon forest. *Ecology* 88: 2259–69.

North, M., J. Innes, and H. Zald. 2007. Comparison of thinning and prescribed fire restoration treatments to Sierran mixed-conifer historic conditions. *Canadian Journal of Forest Research* 37: 331–42.

North, M., B. Oakley, J. Chen, H. Erickson, A. Gray, A. Izzo, D. Johnson, et al. 2002. Vegetation and ecological characteristics of mixed-conifer and red-fir forests at the Teakettle Experimental Forest. In *USFS General Technical Report, PSW-GTR-186.* Berkeley, CA: US Forest Service Pacific Southwest Research Station.

Novick, K. A., R. Oren, P. C. Stoy, M. B. S. Siqueira, and G. G. Katul. 2009. Nocturnal evapotranspiration in eddy-covariance in eddy-covariance records from three colocated ecosystems in the southeastern US: Implications for annual fluxes. *Agricultural and Forest Meteorology* 149: 1491–504.

Novotny, V. 2010. Rain forest conservation a tribal world: Why forest dwellers prefer loggers to conservationists. *Biotropica* 42(5): 546–49.

Novotny, V., and Y. Basset. 2000. Rare species in communities of tropical insect herbivores: Pondering the mystery of singletons. *Oikos* 89, 564–72.

Novotny, V., Y. Basset, S. E. Miller, G. D. Weiblen, B. Bremer, L. Cizek, and P. Drozd. 2002. Low host specificity of herbivorous insects in a tropical forest. *Nature* 416: 841–44.

Novotny, V., P. Drozd, S. E. Miller, M. Kulfan, M. Janda, Y. Basset, and G. D. Weiblen. 2006. Why are there so many species of herbivorous insects in tropical rainforests? *Science* 313: 1115–18.

Novotny, V., S. E. Miller, J. Hulcr, R. A. I. Drew, Y. Basset, M. Janda, G. P. Setliff, et al. 2007. Low beta diversity of herbivorous insects in tropical forests. *Nature* 448, 692–95.

Obregon, A., C. Gehrig-Downie, S. R. Gradstein, R. Rollenbeck, and J. Bendix. 2011. Canopy level fog occurrence in a tropical lowland forest of French Guiana as a prerequisite for high epiphyte diversity. *Agricultural and Forest Meteorology* 151: 290–300.

Odum, H. T. 1970. Rain forest structure and mineral-cycling homeostasis. In *A tropical rain forest,* ed. H. T. Odum and R. F. Pigeon, H-3–52. Oak Ridge, TN: US Atomic Energy Commission.

Odum, H. T., and C. F. Jordan. 1970. Metabolism and evaporation of the lower forest in a giant plastic cylinder. In *A tropical rain forest,* ed. H. T. Odum and R. F. Pigeon, I-165–189. Oak Ridge, TN: US Atomic Energy Commission.

Ogawa, K. 2008. The leaf mass/number trade-off of Kleiman and Aarssen implies constancy of leaf biomass, its density and carbon uptake in forest stands: Scaling up from shoot to stand level. *Journal of Ecology* 96: 188–91.

Oishi, A. C., R. Oren, and P. C. Stoy. 2008. Estimating components of forest evapotranspiration: A footprint approach for scaling sap flux measurements. *Agricultural and Forest Meteorology* 148: 1719–32.

Oliveira-Santos, L. G., M. A. Tortato, and M. E. Graipel. 2008. Activity pattern of Atlantic forest small mammals as revealed by camera traps. *Journal of Tropical Ecology* 24: 563–67.

Olson, J. S. 1963. Energy storage and the balance of producers and decomposers in ecological systems. *Ecology* 44: 322–31.

Ozanne, C. M. P. 2005. Techniques and methods for sampling canopy insects. In *Insect sampling in forest ecosystems*, ed. S. R. Leather, 146–65. Blackwell/J. Wiley and Sons: Malden, MA.

Palace, M., M. Keller, G. P. Asner, S. Hagen, and B. Braswell. 2008. Amazon forest structure from IKONOS satellite data and the automated characterization of forest canopy properties. *Biotropica* 40: 141–50.

Paoletti, M. G., R. A. J. Taylor, B. R. Stinner, D. H. Stinner, and D. H. Benzing. 1991. Diversity of soil fauna in the canopy and forest floor of a Venezuelan cloud forest. *Journal of Tropical Ecology* 7: 373–83.

Parker, G. G. 1995. Structure and microclimate of forest canopies. In *Forest canopies*, ed. M. D. Lowman and N. M. Nadkarni, 73–106. San Diego, CA: Academic Press.

Parker, G. G., and M. J. Brown. 2000. Forest canopy stratification—is it useful? *American Naturalist* 155: 473–84.

Parker, G. G., A. P. Smith, and K. P. Hogan. 1992. Access to the upper forest canopy with a large tower crane. *Bioscience* 42: 664–70.

Parker, T. A., III, B. K. Holst, L. H. Emmons, and J. R. Meyer. 1993. A biological assessment of the Columbia River forest reserve, Toledo District, Belize. *Conservation International. Report* 381.

Parmesan, C., and G. Yohe 2003. A globally coherent fingerprint of climate change impacts across natural systems. *Nature* 421: 37–42.

Parton, W. J., J. M. O. Scurlock, D. S. Ojima, T. G. Gilmanov, R. J. Scholes, D. S. Schimel, T. Kirchner, et al. 1993. Observations and modeling of biomass and soil organic matter dynamics for the grassland biome worldwide. *Global Biogeochemical Cycles* 7: 785–809.

Pendall, E., S. Bridgham, P. J. Hanson, B. Hungate, D. W. Kicklighter, D. W. Johnson, B. E. Law, et al. 2004. Below-ground process responses to elevated CO_2 and temperature: A discussion of observations, measurement methods and models. *New Phytologist* 162: 311–22.

Perrings, C., S. Naeem, F. Ahrestani, D. E. Bunker, P. Burkill, G. Canziani, T. Elmqvist, et al. 2010. Ecosystem services for 2020. *Science* 330: 323–24.

Perry, D. R. 1986. *Life above the jungle floor*. New York: Simon and Schuster.

———. 1995. Tourism, economics, and the canopy: The perspective of one canopy biologist. In *Forest canopies*, ed. M. D. Lowman and N. M. Nadkarni, 605–8. San Diego, CA: Academic Press.

Perry, D. A., R. Oren, and S. C. Hart. 2008. *Forest ecosystems*. 2nd edition. Baltimore, MD: Johns Hopkins University Press.

Pike, L. H., R. A. Rydell, and W. C. Denison. 1977. A 400–year-old Douglas-fir tree and its epiphytes: Biomass, surface area, and their distributions. *Canadian Journal of Forest Research* 7: 680–99.

Pittendrigh, C. 1948. The bromeliad-*Anopheles*-malaria complex in Trinidad. I. The bromeliad flora. *Evolution* 2: 58–89.

Porder, S., G. P. Asner, and P. M. Vitousek. 2005. Ground-based and remotely sensed nutrient availability across a tropical landscape. *Proceedings of the National Academy of Sciences, USA* 102: 10909–12.

Prescott, C. E. 2002. The influence of the forest canopy on nutrient cycling. *Tree Physiology* 22: 1193–200.

Pressley, S., B. Lamb, H. Westberg, and C. Vogel. 2006. Relationships among canopy level energy fluxes and isoprene flux derived from long-term, seasonal eddy covariance measurements over a hardwood forest. *Agricultural and Forest Meteorology* 136: 188–202.

Progar, R. A., T. D. Schowalter, and T. Work. 1999. Arboreal invertebrate responses to varying levels and patterns of green-tree retention in northwestern forests. *Northwest Science* 73 (Special issue): 77–86.

Putz, F. E., and H. A. Mooney, eds. 1991. *The biology of vines.* Cambridge: Cambridge University Press.

Putz, F. E., G. G. Parker, and R. M. Archibald. 1984. Mechanical abrasion and intercrown spacing. *American Midland Naturalist* 112: 24–28.

Puyravaud, J-P., P. Davidar, and W. Laurance. 2010. Cryptic destruction of India's native forests. *Conservation Letters* 3: 390–94.

Pypker, T. G., B. J. Bond, T. E. Link, D. Marks, and M. H. Unsworth. 2005. The importance of canopy structure in controlling the interception loss of rainfall: Examples from a young and an old-growth Douglas-fir forest. *Agricultural and Forest Meteorology* 130: 113–29.

Pypker, T. G., M. H. Unsworth, and B. J. Bond. 2006. The role of epiphytes in rainfall interception by forests in the Pacific Northwest. I. Laboratory measurements of water storage. *Canadian Journal of Forest Research* 36: 809–18.

Ramazzotti, G., and Maucci, W. 1983. Il Phylum Tardigrada. III edizion reveduta e aggiornata. *Memorie dell'Istituto Italiano di Idrobiologia* 41: 1–1011.

Raupach, M. R., J. J. Finnigan, and Y. Brunet. 1996. Coherent eddies and turbulence in vegetation canopies: The mixing-layer analogy. *Boundary-Layer Meteorology* 78: 351–82.

Reagan, D. P. 1995. Lizard ecology in the canopy of an island rain forest. In *Forest canopies*, ed. M. D. Lowman and N. M. Nadkarni, 149–64. San Diego, CA: Academic Press.

———. 1996. Anoline lizards. In *The food web of a tropical rain forest*, ed. D. P. Reagan and R. B. Waide, 321–45. Chicago: University of Chicago Press.

Reagan, D. P., and R. B. Waide, eds. *The food web of a tropical rain forest.* Chicago: University of Chicago Press.

Rebertus, A. J. 1988. Crown shyness in a tropical cloud forest. *Biotropica* 20: 338–39.

Reichle, D. E., and D. A. Crossley Jr. 1967. Investigation on heterotrophic productivity in forest insect communities. In *Secondary productivity of terrestrial ecosystems: Principles and methods*, ed. K. Petrusewicz, 563–87. Warszawa, Poland: Państwowe Wydawnictwo Naukowe.

Reichle, D. E., R. A. Goldstein, R. I. Van Hook, and G. J. Dodson. 1973. Analysis of insect consumption in a forest canopy. *Ecology* 54: 1076–84.

Reynolds, B. C., and M. D. Hunter. 2001. Responses of soil respiration, soil nutrients, and litter decomposition to inputs from canopy herbivores. *Soil Biology and Biochemistry* 33: 1641–52.

Richardson, B. A., M. J. Richardson, G. González, A. B. Shiels, and D. S. Srivastava. 2010. A canopy trimming experiment in Puerto Rico: The response of litter invertebrate communities to canopy loss and debris deposition in a tropical forest subject to hurricanes. *Ecosystems* 11: 286–301.

Richardson, B. A., C. Rogers, and M. J. Richardson. 2000. Nutrients, diversity, and community structure of two phytotelm systems in a lower montane forest, Puerto Rico. *Ecological Entomology* 25: 348–56.

Risley, L. S., and D. A. Crossley Jr. 1988. Herbivore-caused greenfall in the southern Appalachians. *Ecology* 69: 1118–27.

Ross, J. 1981. *The radiation regime and architecture of plant stands*. Hague, Netherlands: W. Junk.

Roubik, D. W. 1993. Tropical pollinators in the canopy and understory: Field data and theory for stratum "preferences." *Journal of Insect Behavior* 6: 659–73.

Ruangpanit, N. 1985. Percent crown cover related to water and soil losses in mountainous forest in Thailand. In *Soil erosion and conservation*, ed. S. A. El-Swaify, W. C. Moldenhauer, and A. Lopp, 462–71. Ankeny, IA: Soil Conservation Society of America.

Ryu, Y., O. Sonnentag, T. Nilson, R. Vargas, H. Kobayashi, R. Wenk, and D. D. Baldocchi. 2009. How to quantify tree leaf area index in an open savanna ecosystem: A multi-instrument and multimodel approach. *Agricultural and Forest Meteorology* 150: 63–76.

Sale, P., ed. 2002. *Coral reef fishes: Dynamics and diversity in a complex ecosystem*. San Diego, CA: Academic Press.

Salo, J., R. Kallioloa, I. Häkkinen, Y. Mäkinen, P. Niemelä, M. Puhakka, and P. D. Coley. 1986. River dynamics and the diversity of Amazon lowland forest. *Nature* 322: 254–58.

Šantrůčková, H., M. I. Bird, J. Frouz, V. Šustr, and K. Tajovský. 2000. Natural abundance of ^{13}C in leaf litter as related to feeding activity of soil invertebrates and microbial mineralization. *Soil Biology and Biochemistry* 32: 1793–97.

Schemske, D. W., and N. Brokaw. 1981. Treefalls and the distribution of understory birds in a tropical forest. *Ecology* 62: 938–45.

Scholander, P., E. Bradstreet, E. Hemmingsen, and H. Hammel. 1965. Sap pressure in vascular plants: Negative hydrostatic pressure can be measured in plants. *Science* 148: 339–46.

Schowalter, T. D. 1993. Cone and seed insect phenology in a Douglas-fir seed orchard during three years in western Oregon. *Journal of Economic Entomology* 87: 758–65.

———. 1995. Canopy arthropod communities in relation to forest age and alternative harvest practices in western Oregon. *Forest Ecology and Management* 78: 115–25.

———. 1999. Throughfall volume and chemistry as affected by precipitation volume, sapling size, and canopy herbivory in a regenerating Douglas-fir ecosystem in western Oregon. *Great Basin Naturalist* 59: 79–84.

———. 2008. Insect herbivore responses to management practices in conifer forests in North America. *Journal of Sustainable Forestry* 26: 204–22.

———. 2011. *Insect ecology: An ecosystem approach*. 3rd edition. San Diego, CA: Elsevier/Academic.

———. 2012. Insect responses to major landscape-level disturbance. *Annual Review of Entomology* 57: 1–20.

Schowalter, T. D., and D. A. Crossley, Jr. 1983. Forest canopy arthropods as sodium, potassium, magnesium and calcium pools in forests. *Forest Ecology and Management* 7: 143–48.

Schowalter, T. D., S. J. Fonte, J. Geaghan, and J. Wang. 2012. Effects of manipulated herbivore inputs on nutrient flux and decomposition in a tropical rainforest in Puerto Rico. *Oecologia* 167: 1141–49.

Schowalter, T. D., and L. M. Ganio. 1998. Vertical and seasonal variation in canopy arthropod communities in an old-growth conifer forest in southwestern Washington, USA. *Bulletin of Entomological Research* 88: 633–40.

———. 2003. Diel, seasonal and disturbance-induced variation in invertebrate assemblages. In *Arthropods of tropical forests*, ed. Y. Basset, V. Novotny, S. Miller, and R. Kitching, 315–28. Cambridge: Cambridge University Press.

Schowalter, T. D., D. C. Lightfoot, and W. G. Whitford. 1999. Diversity of arthropod responses to host-plant water stress in a desert ecosystem in southern New Mexico. *American Midland Naturalist* 142: 281–90.

Schowalter, T. D., and M. D. Lowman. 1999. Forest herbivory by insects. In *Ecosystems of the World: Ecosystems of Disturbed Ground*, ed. L. R. Walker, 269–85. Amsterdam: Elsevier.

Schowalter, T. D., T. E. Sabin, S. G. Stafford, and J. M. Sexton. 1991. Phytophage effects on primary production, nutrient turnover, and litter decomposition of young Douglas-fir in western Oregon. *Forest Ecology and Management* 42: 229–43.

Schowalter, T. D., J. W. Webb, and D. A. Crossley, Jr. 1981. Community structure and nutrient content of canopy arthropods in clearcut and uncut forest ecosystems. *Ecology* 62:1010–19.

Schowalter, T. D., Y. L. Zhang, and R. A. Progar. 2005. Canopy arthropod response to density and distribution of green trees retained after partial harvest. *Ecological Applications* 15: 1594–603.

Seastedt, T. R. 1984. The role of microarthropods in decomposition and mineralization processes. *Annual Review of Entomology* 29: 25–46.

Seastedt, T. R., and D. A. Crossley, Jr. 1981. Microarthropod response following cable logging and clear-cutting in the southern Appalachians. *Ecology* 62: 126–35.

Seastedt, T. R., D. A. Crossley Jr., and W. W. Hargrove. 1983. The effects of low-level consumption by canopy arthropods on the growth and nutrient dynamics of black locust and red maple trees in the southern Appalachians. *Ecology* 64: 1040–48.

Seastedt, T. R., and C. M. Tate. 1981. Decomposition rates and nutrient contents of arthropod remains in forest litter. *Ecology* 62: 13–19.

Sexton, J. M., and T. D. Schowalter. 1991. Physical barriers to reduce *Lepesoma lecontei* (Coleoptera: Curculionidae) damage to conelets in a Douglas-fir seed orchard in western Oregon. *Journal of Economic Entomology* 84: 212–14.

Shaw, D. C. 1998. Distribution of larval colonies of *Lophocampa argentata* Packard, the silver spotted tiger moth (Lepidoptera: Arctiidae), in an old-growth Douglas-fir/western hemlock forest canopy, Cascade Mountains, Washington state, USA. *Canadian Field Naturalist* 112: 250–53.

Shaw, D. C., J. Chen, E. A. Freeman, and D. M. Baun. 2005. Spatial and population characteristics of dwarf mistletoe infected trees in an old-growth Douglas-fir—western hemlock forest. *Canadian Journal of Forest Research* 35: 990–1001.

Shiels, A. B., J. K. Zimmerman, D. C. García-Montiel, I. Jonckheere, J. Holm, D. Horton, and N. Brokaw. 2010. Plant responses to simulated hurricane impacts in a subtropical wet forest, Puerto Rico. *Journal of Ecology* 98: 659–73.

Sillett, S. C., and R. Van Pelt. 2007. Trunk reiteration promotes epiphytes and water storage in an old-growth redwood forest canopy. *Ecological Monographs* 77: 335–59.

Simmons, N., and R. S. Voss. 1998. The mammals of Paracau, French Guiana: A Neotropical lowland rain forest fauna. Part I. Bats. *Bulletin of the American Museum of Natural History* 237: 1–219.

Sojka, R. E. 1999. Physical aspects of soils of disturbed ground. In *Ecosystems of disturbed ground,* ed. L. R. Walkerpp. 503–19. Amsterdam: Elsevier.

Solberg, S., M. Dobbertin, G. J. Reinds, H. Lange, K. Andreassen, P. G. Fernandez, A. Hildingsson, and W. de Vries. 2009. Analysis of the impact of changes in atmospheric deposition and climate on forest growth in European monitoring plots: A stand growth approach. *Forest Ecology and Management* 258: 1735–50.

Song, B., J. Chen, and J. Silbernagel. 2004. Three-dimensional canopy structure of an old-growth Douglas-fir forest. *Forest Science* 50: 376–86.

Southwood, T. R. E. 1961. The number of species of insects associated with various trees. *Journal of Animal Ecology* 30: 1–8.

———. 1978. *Ecological methods with particular reference to the study of insect populations.* London: Chapman and Hall.

Stadler, B., B. Michalzik, and T. Müller. 1998. Linking aphid ecology with nutrient fluxes in a coniferous forest. *Ecology* 79: 1514–25.

Stewart, M. M., and L. L. Woolbright. 1996. Amphibians. In *The food web of a tropical rain forest,* ed. D. P. Reagan and R. B. Waide, 273–320. Chicago: University of Chicago Press.

Sthultz, C. M., C. A. Gehring, and T. G. Whitham. 2009. Deadly combination of genes and drought: Increased mortality of herbivore-resistant trees in a foundation species. *Global Change Biology* 15: 1949–61.

Stork, N. E. 1993. How many species are there? *Biodiversity and Conservation* 2: 215–32.

———. Australian tropical forest canopy crane: New tools for new frontiers. *Austral Ecology* 32: 4–9.

Stouffer, P. C., L. N. Naka, and C. Strong. 2009. Twenty years of understorey bird extinctions from Amazonian rain forest fragments: Consistent trends and landscape-mediated dynamics. *Diversity and Distributions* 15: 88:–97.

Stuntz, S., U. Simon, and G. Zotz. 2002. Rainforest air-conditioning: The moderating influence of epiphytes on the microclimate in tropical tree crowns. *International Journal of Biometeorology* 46: 53–59.

Su, H.-B., H. P. Schmid, C. S. Vogel, and P. S. Curtis. 2008. Effects of canopy morphology and thermal stability on mean flow and turbulence statistics observed inside a mixed hardwood forest. *Agricultural and Forest Meteorology* 148: 862–82.

Sutton, S. L. 2001. Alice grows up: Canopy science in transition from wonderland to reality. In *Tropical forest canopies: Ecology and management,* ed. Linsenmair, K. E., A. J. Davis, B. Fiala, and M. R. Speight, 13–23. Dordrecht, Netherlands: Kluwer Academic.

Tarnay, L., A. W. Gertler, R. R. Blank, and G. E. Taylor Jr. 2001. Preliminary measurements of summer nitric acid and ammonia concentrations in the Lake Tahoe Basin air-shed: Implications for dry deposition of atmospheric nitrogen. *Environmental Pollution* 113: 145–53.

Taylor, P., and M. D. Lowman. 1996. Vertical stratification of small mammals in a northern hardwood forest. *Selbyana* 17: 15–21.

Taylor, S. L., and D. A. MacLean. 2009. Legacy of insect defoliators: Increased wind-related mortality two decades after a spruce budworm outbeak. *Forest Science* 55: 256–67.

Terborgh, J. 1985. The vertical component of plant species diversity in temperate and tropical forests. *American Naturalist* 126: 760–76.

Terborgh, J., L. Lopez, P. Nuñez V., M. Rao, G. Shahabuddin, G. Orihuela, M. Riveros,

R. Ascanio, G. H. Adler, T. D. Lambert, and L. Balbas. 2001. Ecological meltdown in predator-free forest fragments. *Science* 294: 1923–26.

Throop, H. L., E. A. Holland, W. J. Parton, D. S. Ojima, and C. A. Keough. 2004. Effects of nitrogen deposition and insect herbivory on patterns of ecosystem-level carbon and nitrogen dynamics: Results from the CENTURY model. *Global Change Biology* 10: 1092–105.

Toledo-Aceves, T., and J. H. D. Wolf. 2008. Germination and establishment of *Tillandsia eizii* (Bromeliaceae) in the canopy of an oak forest in Chiapas, Mexico. *Biotropica* 40: 246–50.

Tollefson, J. 2010. A towering experiment. *Nature* 467: 386–87.

Torres, J. A. 1988. Tropical cyclone effects on insect colonization and abundance in Puerto Rico. *Acta Científica* 2: 40–44.

———. 1992. Lepidoptera outbreaks in response to successional changes after the passage of Hurricane Hugo in Puerto Rico. *Journal of Tropical Ecology* 8: 285–98.

Trenberth, K. E. 1999. Atmospheric moisture recycling: Role of advection and local evaporation. *Journal of Climate* 12: 1368–81.

Treuhaft, R. N., G. P. Asner, and B. E. Law. 2003. Structure-based forest biomass from fusion of radar and hyperspectral observations. *Geophysical Research Letters* 30: 25-1–4.

Treuhaft, R. N., G. P. Asner, B. E. Law, and S. Van Tuyl. 2002. Forest leaf area density profiles from the quantitative fusion of radar and hyperspectral data. *Journal of Geophysical Research* 107: 4568–80.

Treuhaft, R. N., B. D. Chapman, J. R. dos Santos, F. G. Goncalves, L. V. Dutra, P. Graca, and J. B. Drake. 2009. Vegetation profiles in tropical forests from multibaseline interferometric synthetic aperture radar, field, and lidar measurements. *Journal of Geophysical Research-Atmospheres* 114: Article Number D23110.

Treuhaft, R. N., B. E. Law, and G. P. Asner. 2004. Forest attributes from radar interferometric structure and its fusion with optical remote sensing. *BioScience* 54: 561–71.

Trumble, J. T., D. M. Kolodny-Hirsch, and I. P. Ting. 1993. Plant compensation for arthropod herbivory. *Annual Review of Entomology* 38: 93–119.

Turner, D. P., M. Guzy, M. A. Lefsky, W. D. Ritts, S. van Tuyl, and B. E. Law. 2004. Monitoring forest carbon sequestration with remote sensing and carbon cycle modeling. *Environmental Management* 33: 457–66

Turner, D. P., W. D. Ritts, W. B. Cohen, T. K. Maeirsperger, S. T. Gower, A. A. Kirschbaum, S. W. Running, et al. 2005. Site-level evaluation of satellite-based terrestrial gross primary production and net primary production monitoring. *Global Change Biology* 11: 666–84.

Turner, D. P., W. D. Ritts, B. E. Law, W. B. Cohen, Z. Yang, T. Hudiburg, J. L. Campbell, and M. Duane. 2007. Scaling net ecosystem production and net biome production over a heterogeneous region in the western United States. *Biogeosciences* 4: 597–612.

Uriarte, M., C. D. Canham, J. Thompson, J. K. Zimmerman, and N. Brokaw. 2005. Seedling recruitment in a hurricane-driven tropical forest: Light limitation, density-dependence and the spatial distribution of parent trees. *Journal of Ecology* 93: 291–304.

Van Bael, S. A., A. Aiello, A. Valderrama, E. Medianero, M. Samaniego, and S. J. Wright. 2004. General herbivore outbreak following an El Niño-related drought in a lowland Panamanian forest. *Journal of Tropical Ecology* 20: 625–33.

Van Cleve, K., and S. Martin. 1991. *Long-term ecological research in the United States.* 6th edition. Seattle, WA: University of Washington.

Van Cleve, K., and L. A. Viereck. 1981. Forest succession in relation to nutrient cycling in the boreal forest of Alaska. In *Forest Succession: Concepts and Application,* ed. D. C. West, H. H. Shugart, and D. B. Botkin, 185–211. New York: Springer-Verlag.

Van Pelt, R., and N. Nadkarni. 2004. Development of canopy structure in *Pseudotsuga menziesii* forests in the southern Washington cascades. *Forest Science* 50: 326–41.

Van Pelt, R., S. C. Sillett, and N. M. Nadkarni. 2004. Quantifying and visualizing canopy structure in tall forests: Methods and a case study. In *Forest canopies,* 2nd edition, ed. M. D. Lowman and H. B. Rinker, 49–72. San Diego, CA: Elsevier/Academic Press.

Vehvilainen, H., J. Koricheva, and K. Ruohomaki. 2007. Tree species diversity influences herbivore abundance and damage: Meta-analysis of long-term forest experiments. *Oecologia* 152: 287–98.

Visser, J. H. 1986. Host odor perception in phytophagous insects. *Annual Review of Entomology* 31: 121–44.

Vittor, A. Y., R. H. Gilman, J. Tielsch, G. Glass, T. Shields, W. S. Lozano, V. Pinedo-Cancino, and J. A. Patz. 2006. The effect of deforestation on the human-biting rate of *Anopheles darlingi,* the primary vector of falciparum malaria in the Peruvian Amazon. *American Journal of Tropical Medicine and Hygiene* 74: 3–11.

Waide, R. B. 1996. Birds. In *The food web of a tropical rain forest,* ed. D. P. Reagan and R. B. Waide, 363–98. Chicago: University of Chicago Press.

Walker, L. R. 1991. Tree damage and recovery from Hurricane Hugo in Luquillo Experimental Forest, Puerto Rico. *Biotropica* 23: 379–85.

Walker, L. R., D. J. Lodge, N. V. Brokaw, and R. B. Waide. 1991. An introduction to hurricanes in the Caribbean. *Biotropica* 23: 313–16.

Walker, L. R., and M. R. Willig. 1999. An introduction to terrestrial disturbances. In *Ecosystems of the world 16: Ecosystems of disturbed ground,* ed. L. R. Walker, 1–16. Amsterdam: Elsevier.

Wallace, A. R. 1869. *The Malay archipelago.* London: Macmillan and Co.

———. 1878. *Tropical nature and other essays.* London: Macmillan and Co.

Wang, J., T. W. Sammis, A. A. Andales, L. J. Simmons, V. P. Gutschick, and D. R. Miller. 2006. Crop coefficients of open-canopy pecan orchards. *Agricultural Water Management* 88: 253–62.

Waring, R. H., and J. F. Franklin. 1979. Evergreen coniferous forests of the Pacific Northwest. *Science* 204: 1380–86.

Waring, R. H., and S. W. Running. 2007. *Forest ecosystems: Analysis at multiple scales.* 3rd edition. San Diego, CA: Academic Press.

Wassie-Eshete, A. 2007. Ethiopian church forests: Opportunities and challenges for restoration. PhD diss., Wageningen Universitat, The Netherlands.

Watkins, J. E. Jr., C. Cardelús, R. K. Colwell, and R. C. Moran. 2006. Species richness and distribution of ferns along an elevational gradient in Costa Rica. *American Journal of Botany* 93: 73–83.

Watkins, J. E., M. K. Mack, and S. S. Mulkey. 2007. Gametophyte ecology and demography of epiphytic and terrestrial tropical ferns. *American Journal of Botany* 94: 701–8.

Weaver, D. B. 2001. *The encyclopedia of ecotourism.* Wallingford, UK: CABI.

Webb, E. K., G. J. Pearman, and R. Leuning. 1980. Correction of flux measurements for density effects due to heat and water vapor transfer. *Quarterly Journal of the Royal Meteorological Society* 106: 85–100.

Whelan, C.J. 2001. Foliage structure influences foraging of insectivorous forest birds: An experimental study. *Ecology* 82: 219–31.

Whigham, D.F., M.B. Dickinson, and N.V.L. Brokaw. 1999. Background canopy gap and catastrophic wind disturbances in tropical forests. In *Ecosystems of the world 16: Ecosystems of disturbed ground*, ed. L.R. Walker, 223–52. Amsterdam: Elsevier.

White, T., B. Asfaw, Y. Beyene, Y. Haile-Selassie, C.O. Lovejoy, G. Suwa, and G. WoldeGabriel. 2010. Ardipithecus ramidus and the paleobiology of early hominoids. *Science* 326: 75–86.

White, P.S., and S.T.A. Pickett. 1985. Natural disturbance and patch dynamics: An introduction. In *The ecology of natural disturbance and patch dynamics*, ed. S.T.A. Pickett and P.S. White, 3–13. Orlando, FL: Academic Press.

Whitford, W.G., V. Meentemeyer, T.R. Seastedt, K. Cromack Jr., D.A. Crossley Jr., P. Santos, R.L. Todd, and J.B. Waide. 1981. Exceptions to the AET model: Deserts and clear-cut forest. *Ecology* 62: 275–77.

Willig, M.R., and L.R. Walker. 1999. Disturbance in terrestrial ecosystems: Salient themes, synthesis, and future directions. In *Ecosystems of the world: Ecosystems of disturbed ground*, ed. L.R. Walker, 747–67. Amsterdam: Elsevier.

Wilson, E.O. 1992. *The diversity of life*. Cambridge, MA: Harvard University Press.

Winchester, N.N. 2006. Ancient temperate rain forest research in British Columbia. *Canadian Entomologist* 138: 72–83.

Windsor, D.M. 1990. *Climate and moisture variability in a tropical forest: Long-term records from Barro Colorado island, Panamá*. Washington, DC: Smithsonian Institute Press.

Wire Rope Technical Board (WRTB). 1993. *Wire rope users manual.* 3rd edition. Woodstock, MD: WRTB.

Wullaert, H., T. Pohlert, J. Boy, C. Valarezo, and W. Wilcke. 2009. Spatial throughfall heterogeneity in a montane rain forest in Ecuador: Extent, spatial stability and drivers. *Journal of Hydrology* 377: 71–79.

Wullschleger, S.D., F.C. Meinzer, and R.A. Vertessy. 1998. A review of whole-plant water use studies in trees. *Tree Physiology* 18: 499–512.

Wunderle, J.M., Jr., L.M.P. Henriques, and M.R. Willig. 2006. Short-term responses of birds to forest gaps and understory: An assessment of reduced-impact logging in a lowland Amazon forest. *Biotropica* 38: 235–55.

Wunderle, J.M., Jr., and R.B. Waide. 1993. Distribution of overwintering Nearctic migrants in the Bahamas and Greater Antilles. *Condor* 95: 904–33.

Xiao, J., Q. Zhuang, D.D. Baldocchi, B.E. Law, A.D. Richardson, J. Chen, R. Oren, et al. 2008. Estimation of net ecosystem carbon exchange for the coterminous United States by combining MODIS and AmeriFlux data. *Agricultural and Forest Meteorology* 148: 1827–47.

Xu, C.-Y., and K.L. Griffin. 2008. Scaling foliar respiration to the stand level throughout the growing season in a Quercus rubra forest. *Tree Physiology* 28: 637–46.

Yanoviak, S.P. 1999. Effects of leaf litter species on macroinvertebrate community properties and mosquito yield in Neotropical tree hole microcosms. *Oecologia* 120: 147–55.

———. 2001. Predation, resource availability, and community structure in Neotropical water-filled tree holes. *Oecologia* 126: 125–33.

Yanoviak, S. P., N. M. Nadkarni, and R. Solano J. 2007. Arthropod assemblages in epiphyte mats of Costa Rican cloud forests. *Biotropica* 39: 202–10.

Yokelson, R. J., S. P. Urbanski, E. L. Atlas, D. W. Toohey, E. C. Alvarado, J. D. Crounse, P. O. Wennberg, et al. 2007. Emissions from forest fires near Mexico City. *Atmospheric Chemistry and Physics* 7: 5569–84.

Zhang, J., Y. Hu, X. Xiao, P. Chen, S. Han, G. Song, and G. Yu. 2009. Satellite-based estimation of evapotranspiration of an old-growth temperate mixed forest. *Agricultural and Forest Meteorology* 149: 976–84.

Zimmerman, J. K., E. M. Everham III, R. B. Waide, D. J. Lodge, C. M. Taylor, and N. V. L. Brokaw. 1994. Responses of tree species to hurricane winds in subtropical wet forest in Puerto Rico: Implications for tropical tree life histories. *Journal of Ecology* 82: 911–22.

Zimmerman, J. K., and I. C. Olmstead. 1992. Host tree utilization by vascular epiphytes in a seasonally inundated forest (Tintal) in Mexico. *Biotropica* 24: 402–7.

Zotz, G. 1997. Substrate use of three epiphytic bromeliads. *Ecography* 20: 264–70.

———. 2005. Differences in vital demographic rates in three populations of the epiphytic bromeliad, *Werauhia sanguinolenta*. *Acta Oecologia* 28: 306–12.

AUTHOR INDEX

Van Cleve, K., 17
Van Pelt, R., 50, 54, 60
Vehvilainen, H., 63, 172
Visser, J., 126
Vittor, A., 135

Waide, R., 85, 99, 100
Walker, L., 16, 63, 113
Wallace, A., 3, 43, 154
Wang, J., 60
Waring, R., 11, 14, 22, 25
Wassie-Eshete, A., 1, 160
Watkins, J., 76, 77
Weaver, D. B., 161
Webb, E., 115
Whelan, C., 54, 64, 100, 172
Whigham, D., 16
White, P., 16
White, T., 2, 7, 41, 57, 154

Whitford, W., 135
Willig, M., 16
Wilson, E., 2, 27, 34, 154
Winchester, N., 54, 58, 73
Windsor, D., 113
Wire Rope Technical Board (WRTB), 31
Wullaert, H., 102
Wullschleger, S., 136, 137
Wunderle, J., 59, 63, 90

Xiao, J., 121
Xu, C., 170

Yanoviak, S., 58, 65, 70, 172
Yokelson, R., 125

Zhang, J., 123
Zimmerman, J. K., 55, 75
Zotz, G., 75, 77

SUBJECT INDEX

bait, 43–44, 84–85, 90

balloon, 5, 37, 46

barrier, 102, 126, 161

behavior, 54, 86, 100

Belize, 40, 164

Berlese funnel, 84

binoculars, 4, 43, 74, 169

biodiversity, 2–3, 6–8, 12, 17, 28, 32, 34, 41–45, 47, 50, 65, 68, 154, 158, 160, 166, 168, 173
  measurement of, 33–39, 100, 104, 157

biogenic gas, 110

biomass, 11, 60–61, 63, 74, 129, 132–34, 148, 150, 172

bird, 2–4, 6–7, 16, 29, 33, 41, 54–55, 57, 64, 68–70, 73, 85, 90, 99–100, 135, 161, 166, 171–72

bole, 49, 54, 59–60, 65, 74, 80, 85, 90–91, 101–2, 115, 117, 123, 133, 137–38, 140, 144, 150
  breakage, 51, 113
  coordinates, 60
  decay, 58
  height, 53
  mapping, 21
  orientation, 53

boom, 5, 45–46, 120

boreal forest, 12, 14, 16, 23, 51, 52–54, 118–19, 170

Borneo, 59

boundary layer, 24, 69, 137

branch, 2–6, 15, 28, 30, 43–44, 49–50, 57, 61, 63, 70–71, 76, 78, 85–86, 89, 92, 94, 96–97, 99, 102, 104, 110, 114, 117, 134–35, 172
  angle, 31, 53, 56, 59, 61, 65, 77, 97
  bagging, 80–82, 84, 91, 171
  beating, 80
  breakage, 51, 55, 58, 135, 141
  density, 54, 56, 75, 100, 106, 113, 172
  growth, 16
  height, 54, 59–60, 75, 77, 103, 113
  length, 53, 59–60, 74–75, 80
  network, 43, 54, 59, 61, 106, 112, 133
  orientation, 53–54, 59, 65, 100, 106, 123, 172
  size, 54, 58, 69, 74, 77, 144
  spacing, 54, 59, 64, 103, 172

Brazil, 4, 33, 43–44, 85, 122, 158

bridge, 5, 30

British Columbia, 58

British Guiana, 43

bromeliad, 2, 58, 69, 75

butterfly farm/garden, 164

calibration, 19, 61–62, 119, 136

camera, 19–20, 28, 84–86, 88, 89–90, 138

Cameroon, 40, 46, 164

canopy architecture, 29–31, 53–54, 59–60, 65, 100, 149, 172

canopy-atmosphere interaction, 9, 28, 50, 109–29
  manipulation of, 127–28
  measurement of, 115–26

canopy-atmosphere interface, 6, 28, 46

canopy biomass, 63, 133, 172

canopy cooling, 111–12, 132

canopy cover, 22, 28, 46–47, 50, 56, 60, 63, 91, 106, 109–10, 112, 114, 127, 129, 131–32, 136, 144, 153

canopy density, 90, 112

canopy depth, 113, 132, 169, 171

canopy–forest floor interaction, 9, 56, 131–51
  manipulation of, 144–50
  measurement of, 135–43

canopy height, 19, 22–23, 50–51, 60, 64, 84, 101, 110, 112, 172

canopy mapping, 20–23, 30, 59–60, 65, 96–97

canopy microclimate, 28, 114–15, 128, 171

canopy opening, 16, 63, 103, 126, 132, 136, 144–45, 151, 172

canopy processes, 9, 29–30, 40–41, 47, 59, 73–74, 105, 157, 171
  measurement of, 91–99

canopy roughness, 112

canopy shyness, 15, 55

canopy strata, 36, 51, 57–58, 62, 64, 74, 100–101, 172

canopy structure, 14, 16–19, 22, 68, 79, 99,·101, 106, 115, 125–27, 133, 135, 143–44, 169, 171
  manipulation of, 127–28
  measurement of, 19–22, 49–65

canopy surface area, 110, 132, 134

canopy topography, 16–17, 19, 23, 50–51, 65

capillary action, 133

carbohydrate, 73, 110, 131, 134

carbon, 61, 74, 158

carbon credit, 158–59

carbon dioxide, 118, 120, 126–29

carbon exchange, 51, 58, 102, 109–10, 121, 127, 129
    measurement of, 117–22
carbon flux, 22–23, 49, 62, 65, 106, 127, 132, 134, 144, 148, 171, 173
carbon isotope, 140, 143
carbon monoxide, 111
carbon storage/sequestration, 1, 7, 9, 11, 19, 23, 28, 41, 110, 118–19, 129, 157, 160
carbon uptake, 118–19
Caribbean, 17
cell cavitation, 114, 117, 133, 136
cellulose, 73
census, 21, 33, 78, 85
Chile, 135
chlorophyll, 14, 92, 138
chromatograph, 115, 125
chronosequence, 118–19
citric acid, 125
climate, 6, 8, 12, 14, 22–23, 65, 106, 118, 127, 129, 132, 159, 171, 173
    change, 3, 6, 22–23, 28, 39, 47, 63, 68, 94, 97, 99, 103, 157–59, 165–68
    measurement of, 115
    modification, 111–13
    stabilization/control, 7, 41, 157
climbing, 19, 27, 32–33, 41, 44, 59, 74, 82, 86, 89, 103, 169
    equipment, 4, 5, 44–45, 71, 161
cloud forest, 13, 15, 33
cloud formation, 110, 112, 129
coarse woody debris, 53, 113, 118, 134
colonization, 58, 65
community, 2, 7, 28, 30, 36, 53, 80, 93, 110, 114
    analysis, 76
    assembly, 105
    composition, 16, 106
    development, 58, 65, 172
    responses, 105–6
    structure, 59, 99, 105
competition, 14, 16–17, 53, 61
competitor, 16
Congo, 158
conifer forest, 13, 14, 24–25, 53–54, 56, 58, 63, 90–92, 137, 140, 170, 172
conservation, 3, 7, 9, 11, 19, 28, 30, 33, 41, 46–47, 153–54, 157, 160–61, 163–65, 167–69, 172–73

consumer/consumption, 2, 39–40, 85, 94, 96, 146, 148
    rate of, 40, 91
convection, 112
corridor, 55
Costa Rica, 4, 33, 42, 44, 59, 76, 156–57
crane, 5–6, 13, 19, 28, 30, 37–38, 40, 42, 46–47, 59, 74, 82, 84, 91, 100, 103, 170
crown, 4–5, 33, 40–41, 45–47, 50, 57, 60–61, 64, 83–84, 90, 94, 97, 102–4, 113–14, 124, 128, 149–50, 154, 171–72
    architecture, 59, 172
    area, 21, 60
    damage, 16, 55, 60
    death, 53
    depth, 53–54
    emergent, 16, 18, 50
    form/shape, 19–20, 23, 51–53, 58,
    gap, 15
    height, 19, 69, 82, 94
    mapping, 20, 59, 65
    perimeter, 9, 44, 59–61, 74, 123
    position, 21, 55, 74
    shyness, 15, 55
    spacing, 19, 50, 54–56, 60, 65
    structure, 59–60
cultural heritage, 1, 7, 11, 41
cyclonic storm, 16, 113

damping ratio, 112
data, 22–23, 28, 33–34, 36–39, 43, 46, 58, 60–62, 65, 75, 86, 100, 102, 106, 119–21, 123, 157, 168, 170–71
    analysis, 19, 22, 33, 37, 105, 172
    collection, 4, 7, 19–20, 22, 36, 41, 44–45, 67, 70, 74, 76, 85, 109, 131, 166
    logger, 124, 136–37, 150
    pooling, 103–4
    pretreatment, 63, 172
    quality, 23, 80, 104
    recording, 71–72, 138
    requirements, 24
    scaling, 169–70
    transformation, 104
dead trees, 51, 53
decay constant, 92, 142–43
decay rate, 73, 142–43

decomposition, 40, 70, 73, 110, 131–32, 134
    manipulation of, 102, 144
    measurement of, 92, 136, 140, 142–43, 145,
        149, 151
    rate, 25, 92, 135, 144
defoliation, 103, 117
deforestation, 8, 11, 23, 28, 109, 112, 127, 135, 153,
    158–59, 167, 171, 173
deposition, 22, 110–11, 126, 145
    measurement of, 125, 129
desiccation, 114, 133
detritivore, 68–70, 73, 100, 102, 104, 143
deuterium, 137
disease transmission, 135
disturbance, 12, 14, 15, 16–17, 22, 23, 50, 53, 60,
    62–63, 65, 68, 74, 85, 99–100, 105, 110–
    15, 118–19, 127, 129, 131–34, 138, 143–44,
    171
DNA sequence, 154–55
domain reflectometry, 136, 149
double rope techniques (DRT), 4, 44
drainage, 15, 120, 133
drought, 14–16, 53, 68, 75, 92, 99, 112–14, 117,
    129, 133, 136, 138, 144
DRT. *See* double rope techniques (DRT)

economic, 41, 47, 165
    driver, 168
    policy, 9
    products, 9, 154
    security, 160
    tradeoff, 161, 164
    values, 7
ecosystem, 2, 3, 6, 8, 11, 28–29, 39, 63, 65, 99–
    100, 111, 134, 149, 154–55, 164, 166, 172
    boundaries, 18
    degradation, 7
    dynamics, 149
    model, 119
    processes, 50, 65, 106, 110, 114, 118–21, 143–
        44, 146–48, 150
    production/productivity, 51
    services, 1, 7, 9, 18, 22, 41, 110, 153–54, 157–58,
        160–61, 163, 167–69, 173
    stability, 49
    type, 147
ecotone, 18, 23, 114

ecotourism, 5, 30–31, 45–46, 154, 161, 164, 166,
    172–73
eddy
    covariance, 110, 115, 129, 137, 143, 150
    measurement, 116, 118–23,
    turbulence, 55, 126
edge, 78, 114, 126, 134
education, 9, 154–55, 161, 165, 166–69, 172–73
electrical grid, 144
electrical signal, 138
elevation, 15, 18, 40, 53, 76, 99
energy, 1, 2, 49, 133, 137, 159
    capture, 6, 7, 8, 24, 112
    flux, 17, 106, 110, 118, 120, 157
    use, 56, 65
environment, 8, 11, 16–18, 28–29, 37, 39, 43,
    50–51, 53, 56–57, 62, 65, 68–70, 73, 82,
    92, 103, 111–12, 114, 153, 161, 165–66, 168,
    170–71, 173
    change in, 3, 63, 99–100, 105–6, 131, 172
    health of, 2
epiphyte, 2–4, 30, 33, 40–41, 44–46, 51, 53–56,
    58, 61–62, 65, 68–70, 80, 84, 100–101,
    104, 106, 110, 114, 135, 154, 157, 164
    measuring abundance of, 74–79
equatorial convergence zone, 14
equipment, 4, 20, 41, 59, 71, 84, 110, 136
    monitoring, 60, 67, 85, 86, 100
    placement of, 59, 67, 74, 115, 117, 121
erosion, 56, 111, 131, 133, 140, 150
Ethiopia, 2, 160–61, 163
EuroFlux network, 171
evaporation, 15, 56, 80, 117, 123–24, 132–33, 150
evapotranspiration, 11, 56, 110, 112–13, 121, 129,
    133, 137
    measurement of, 62, 123–25
    rate of, 111, 132–33, 171
exclusion/inclusion cage, 91, 99, 147
experimental design, 4, 7, 50, 63, 94, 106, 148
experimental methods, 100–102
experimental manipulation, 23, 62–65, 100–
    102, 106, 110, 127–28, 132, 144–51, 169,
    171–72

FACE. *See* free-air CO$_2$ enrichment (FACE)
feces/frass, 73, 91, 102, 134–36, 139–42, 145–48,
    151

flux, 23, 148
 isotope, 138, 140, 143, 148
 retrieval, 56, 73
nitrous oxide, 140
normalized difference vegetation index (NDVI),
 61, 92
North American Carbon Program, 119
nutrient, 2, 56, 69, 131, 135, 146–47
 availability, 56, 61–62, 73, 146, 171
 capture/input, 49, 53, 111, 122, 150
 concentration, 73, 77, 125, 136, 138, 142–43,
 150, 160
 cycling, 2, 8, 40–41, 110, 157
 deposition, 110, 125–26
 distribution, 68
 export, 131, 139, 143
 flux, 49, 73, 101–2, 106, 111, 121, 132–41, 144–
 45, 150–51, 171
 retrieval, 14, 56, 73
 storage, 11, 74, 110
 uptake, 131, 138

Operation Drake, 4
oxidation, 111
oxygen, 6, 41, 113
ozone, 111

paired-plot design, 63, 127
Panama, 5, 6, 20–21, 34–35, 37, 39–40, 43, 45–
 46, 64, 101, 104, 166
Papua New Guinea, 2, 33, 155
paraecologist, 154–57
parasite, 4, 43, 55, 68–69, 74, 155
parataxonomist, 28, 37, 156
path length, 53, 133
Peru, 6, 15, 31, 33, 40, 161
phenology, 4–5, 40–41, 44, 99
pheromone, 111
phloem, 134
phosphate, 140
phosphorus, 74
photography, 38, 87, 97
 aerial, 20, 60
 digital, 19–20, 60
 fish eye, 60
 flash, 99
 hemispheric, 60

infrared, 86
 methods/technique, 60
 video, 90
photoionization detector, 115
photosynthate, 140, 150
photosynthesis, 5, 14, 40–41, 47, 58, 69, 110,
 129, 133–34, 140, 144, 170–71
 measurement of, 117–19, 125
 rate of, 56, 61, 73, 99
photosynthetic efficiency, 50, 56
phytotelm, 2, 110
plantation, 8, 119, 153, 171
plant defense, 7
plant growth, 14, 132
plant stress, 16, 25, 92, 111, 114, 133, 136–38
platform, 5, 22, 30,·32, 40, 43, 46, 53, 74, 100,
 133, 138, 164–65, 169–70, 173
plot size, 35–36, 61, 63, 65, 106, 124, 144, 155, 157
point count, 85
pole, 5, 33, 44–45, 59, 82, 99
policy maker, 159, 165–66
pollination, 5, 41, 45, 57, 163
pollinator, 4–6, 8, 35–36, 39, 41, 44, 46, 57, 64,
 68, 73, 101, 160–61, 172
pollutant/pollution, 63, 68, 111, 158–59
population, 3, 6, 16, 33, 47, 74
 density, 80, 85, 90–91, 147
 dynamics, 77
 growth, 98, 158
 size, 104
positive feedback, 112, 132
potassium, 113, 147
potentiometer, 19
power source, 74, 90, 115
precipitation, 11, 14–15, 24, 54, 55, 91, 109–13, 115,
 121, 124, 129, 132, 138–41, 144–45, 169, 173
 interception of, 24, 51, 54, 56, 110, 114, 127–
 29, 133, 150
 measurement of, 122–23
predation, 74, 91, 99, 102, 110, 155
predator, 9, 16, 33, 35–36, 58, 65, 69–70, 90,
 102, 104, 135, 172
 rate of, 74
 measurement of, 98–99
pressure bomb, 136
primary production, 14, 17–18, 61, 117, 129, 134,
 170–71

probe, 137
   moisture, 74, 115, 136
   temperature, 74, 115, 125, 150
protein, 73
protractor, 58, 77
pseudoreplication, 103, 127
Puerto Rico, 33, 40, 102, 113, 117

radiometer, 60, 138
raft, 5, 30, 31, 36–39, 46–47
raindrop impact, 56, 140
rainforest, 2–6, 14–15, 31, 34–35, 39–41, 43–45,
     49, 55, 63, 84–86, 90, 93–94, 97, 104, 113,
     117, 135
rain gauge, 74, 115, 138
rare earth, 143
reflectance, 19, 74, 92, 138
refugia, 161
relative humidity, 51, 54, 58, 68–69, 74, 109, 111–
     12, 114–15, 123, 132–33, 137, 171
   measurement of, 124, 125
religious leaders, 160, 163
remote sensing, 19–20, 22–23, 46, 60–62, 65,
     67, 74, 92, 106, 110, 119, 137, 170
   raster method, 22
   vector method, 22
repeated measures, 104
repellent/toxic chemical, 80, 89, 92, 101
replication, 4–5, 23, 36, 44, 50, 62, 63, 65, 67,
     76, 94, 96–97, 102–4, 106, 115, 127, 136,
     138, 143–46, 151, 172
reptile, 2, 33, 57, 69, 85
respiration, 24–25, 99, 110, 117, 129, 131–32, 135,
     144–45, 170
   measurement of, 118–21, 142–43
restoration, 19, 37, 63, 153, 161, 172
rhizosphere, 131, 134, 140, 142, 150
road, 84
root, 3, 15–16, 24–25, 30, 54, 62, 112–13, 115, 119,
     134, 138, 140, 142, 149–50
ropes, 3–5, 29, 30, 32–33, 41–45, 47, 59, 70, 80,
     82, 161, 169

safety, 5, 27, 30–31, 44–45, 60, 64, 68, 71, 75, 172
Samoa, 164
sample, 22, 29, 37, 39, 45, 74, 86, 91–92, 115,
     122, 125, 138, 143, 147–48, 151, 156, 161

   independence of, 62–63, 103–4, 106
   pooling, 103
   repeated, 104
   unit, 49–50, 61–62, 65, 103, 106
sampling, 33, 36, 40–41, 44–45, 47, 62, 67, 120,
     155
   efficiency, 72, 80–85, 93, 99–100, 104
   effort, 8, 11, 32, 82–83, 90, 100
   long term, 50, 75–79, 94, 99
   method, 74–106, 90
   mobility, 82
   random, 59, 103
sap flux, 123–25, 136, 150
   measurement of, 137
saprophage, 53
satellite imagery, 20, 28, 46, 123, 159, 166, 169
scaffold, 5, 46
scale, 1, 3, 16–17, 20, 23, 29, 34, 36, 39, 41, 47,
     61–63, 65, 75, 82, 91, 99, 113, 120, 122, 124,
     128, 137, 144, 146, 149–50, 155, 157–58, 160,
     168–71
scaling, 21, 61–62, 94, 117, 121, 169–70
scorch height, 16, 117
Scotland, 5
sediment, 58, 111, 131, 140
sediment collector, 140
seed, 57, 60–61, 99, 101, 160
   dispersal, 17, 57, 90, 113
   germination, 15
seedling, 99
   mortality, 47, 114
sensible heat loss, 112
sensor array, 74, 115
   placement, 49, 59, 106, 115, 120
shade tolerance, 16, 54
shading, 1, 4, 11, 16, 28, 40, 43, 112, 129, 132, 135,
     150, 160, 164, 167
single rope techniques (SRT), 4, 32–33, 39–41,
     44–46
sleeve cage, 91, 99
slingshot, 3, 33, 44
smoke, 126
snow, 24, 148
   storm, 16
soda lime, 142–43
sodium carbonate, 125
sodium hydroxide, 142

soil, 2–3, 7, 22, 41, 63, 118, 146, 160
    accumulation/development, 54, 58, 65, 68–69, 74, 77, 111, 150
    aggregate, 134
    arboreal, 53, 58, 65, 69, 84, 101
    community, 2, 28, 34, 53, 58, 69, 84, 104, 134, 140
    desiccation, 133
    disturbance, 15, 113
    erosion/displacement, 111, 133, 140, 150
    fertility, 18, 132, 135, 144–45
    habitat, 64, 101, 172
    hypoxia, 113
    manipulation of, 101
    moisture/water, 11, 15, 25, 53, 54, 64, 101, 111, 113–14, 121, 124–25, 131, 133–34, 136–37, 144, 149, 150, 172
    nutrients, 24, 79, 131, 134, 138–41, 145–48
    respiration, 25, 119, 132, 135, 143–45
    temperature, 56, 117, 132, 135–36, 144, 150
    type, 12, 14–15
    warming, 132–33, 144, 151
solar
    energy, 112
    exposure, 14–15, 51, 53–54, 132
    power, 74, 89, 115–16, 169
    radiation, 23, 53
species, 7, 9, 11–12, 14, 19, 29, 43, 47, 55, 62, 64, 68, 71, 82, 86, 96–97, 99–103, 124, 135, 137, 143, 146, 172
    abundance, 3, 76–77, 84–85, 105–6, 127, 171
    composition, 12, 14, 19, 21, 23, 46, 51, 53–54, 56–58, 61
    distribution, 16–17, 20, 77, 84
    diversity, 3, 7, 14, 21, 45, 69, 94, 154
    extinction, 6–8, 15–16
    identification, 6, 8
    indicator, 161
    interactions, 111, 114, 154–55
    keystone, 41
    pioneer, 50, 113
    pool, 85
    replacement, 16–17, 105, 114
    richness, 2, 6, 33–34, 36, 69, 73, 75–78, 85, 104–6, 154, 156
    turnover, 15–16, 35, 36, 105
spectrometry, 60, 74, 106, 125, 148

spectrophotometry, 74, 121, 125, 136, 142
spectroradiometry, 19, 138
spot map, 85
sprout/sprouting, 62, 113
SRT. *See* single rope techniques (SRT)
stable isotope, 136–40, 142–43, 147–48, 150–51
stand, 6, 15, 21, 28, 47, 50, 53, 62, 94, 118–20, 126, 139, 146, 153, 160, 172
statistical, 19, 22,
    analysis, 50, 63, 76, 103–6, 115, 127
    comparison, 18, 50, 151, 172
    independence, 62, 106
    multivariate, 37
    power, 76, 102
stem, 6, 20, 24–25, 28, 61, 63, 92, 117, 128, 149, 155, 172
STEM (science, technology, engineering, and mathematics) education, 165
stemflow, 123, 133–34, 139, 150
    measurement of, 123, 139
stewardship, 9, 160, 168
stomata, 109
    closure, 24, 113
    conductance, 24, 123, 137
storm, 16, 18, 53, 55, 68, 99, 112, 117, 129, 134–35
    cyclonic, 16, 17, 113
    ice, 16
    snow, 16
strain transducer, 117
stress index, 92
subcanopy, 19, 50, 55, 89, 102, 113–14
subsistence agriculture, 161
succession, 17, 19, 50, 114–15, 118, 121
    early, 113
    late, 51
Sulawasi, 4
surface area, 110, 132, 134
sustainability, 5, 28, 49, 154, 161, 164–65, 167
swamp, 15
sway, 15, 113, 115
    frequency, 112–13
    measurement of, 117
Switzerland, 6, 46

telescope, 4, 43, 138
temperate forest, 5, 12–16, 23, 34, 39, 53, 57, 73, 170

COMPOSITOR: Scribe Inc.

TEXT: 9.5/14 Scala

DISPLAY: Scala Sans

PREPRESS: Embassy Graphics

PRINTER AND BINDER: QuaLibre